K-Theory

M. F. ATIYAH
Oxford University, England

Notes by
D.W. ANDERSON

CRC Press
Taylor & Francis Group
Boca Raton London New York

CRC Press is an imprint of the
Taylor & Francis Group, an **informa** business
A CHAPMAN & HALL BOOK

Originally published in 1967 as part of the Mathematics
Lecture Note Series by W. A. Benjamin, Inc.

Work for these notes was partially supported by NSF Grant CP-1217

Published 1989 by Westview Press

Published 2018 by CRC Press
Taylor & Francis Group
6000 Broken Sound Parkway NW, Suite 300
Boca Raton, FL 33487-2742

ISBN 13: 978-0-201-40792-1 (pbk)

Library of Congress Cataloging-in-Publication Data

Atiyah, Michael Francis, 1929-
K-theory : lectures / by M.F. Atiyah : notes by D. W, Anderson
p. cm. --(Advanced book classics series)
"Fall 1964,"
Reprint. Originally published: New York : W.A. Benjamin, 1967.
Includes bibliographical references.
1. K-theory. I. Anderson, D. W. II. Title III. Series.
QAC12.33.A85 1988 514'.23--dc19 88-22327

ISBN 0-201-09394-4 (Hardcover) ISBN 0-201-40792-2 (Paperback)

K-Theory

This unique image was created with special-effects photography. Photographs of a broken road, an office building, and a rusted object were superimposed to achieve the effect of a faceted pyramid on a futuristic plain. It originally appeared in a slide show called "Fossils of the Cyborg: From the Ancient to the Future," produced by Synapse Productions, San Francisco. Because this image evokes a fusion of classicism and dynamism, the future and the past, it was chosen as the logo for the Advanced Book Classics series.

Publisher's Foreword

"Advanced Book Classics" is a reprint series which has come into being as a direct result of public demand for the individual volumes in this program. That was our initial criterion for launching the series. Additional criteria for selection of a book's inclusion in the series include:

- Its intrinsic value for the current scholarly buyer. It is not enough for the book to have some historic significance, but rather it must have a timeless quality attached to its content, as well. In a word, "uniqueness."
- The book's global appeal. A survey of our international markets revealed that readers of these volumes comprise a boundaryless, worldwide audience.
- The copyright date and imprint status of the book. Titles in the program are frequently fifteen to twenty years old. Many have gone out of print, some are about to go out of print. Our aim is to sustain the lifespan of these very special volumes.

We have devised an attractive design and trim-size for the "ABC" titles, giving the series a striking appearance, while lending the individual titles unifying identity as part of the "Advanced Book Classics" program. Since "classic" books demand a long-lasting binding, we have made them available in hardcover at an affordable price. We envision them being purchased by individuals for reference and research use, and for personal and public libraries. We also foresee their use as primary and recommended course materials for university level courses in the appropriate subject area.

The "Advanced Book Classics" program is not static. Titles will continue to be added to the series in ensuing years as works meet the criteria for inclusion which we've imposed. As the series grows, we naturally anticipate our book buying audience to grow with it. We welcome your support and your suggestions concerning future volumes in the program and invite you to communicate directly with us.

Advanced Book Classics

1989 Reissues

V.I. Arnold and A. Avez, *Ergodic Problems of Classical Mechanics*

E. Artin and J. Tate, *Class Field Theory*

Michael F. Atiyah, *K-Theory*

David Bohm, *The Special Theory of Relativity*

Ronald C. Davidson, *Theory of Nonneutral Plasmas*

P.G. de Gennes, *Superconductivity of Metals and Alloys*

Bernard d'Espagnat, *Conceptual Foundations of Quantum Mechanics, 2nd Edition*

Richard Feynman, *Photon-Hadron Interactions*

William Fulton, *Algebraic Curves: An Introduction to Algebraic Geometry*

Kurt Gottfried, *Quantum Mechanics*

Leo Kadanoff and Gordon Baym, *Quantum Statistical Mechanics*

I.M. Khalatnikov, *An Introduction to the Theory of Superfluidity*

George W. Mackey, *Unitary Group Representations in Physics, Probability and Number Theory*

A. B. Migdal, *Qualitative Methods in Quantum Theory*

Phillipe Nozières and David Pines, *The Theory of Quantum Liquids, Volume II* - new material, 1989 copyright

David Pines and Phillipe Nozières, *The Theory of Quantum Liquids, Volume I: Normal Fermi Liquids*

David Ruelle, *Statistical Mechanics: Rigorous Results*

Julian Schwinger, *Particles, Source and Fields, Volume I*

Julian Schwinger, *Particles, Sources and Fields, Volume II*

Julian Schwinger, *Particles, Sources and Fields, Volume III* - new material, 1989 copyright

Jean-Pierre Serre, *Abelian ℓ-Adic Representations and Elliptic Curves*

R.F. Streater and A.S. Wightman, *PCT Spin and Statistics and All That*

René Thom, *Structural Stability and Morphogenesis*

Vita

Sir Michael F. Atiyah

Born in London on April 22, 1929, Sir Michael Atiyah is a Royal Society Research Professor and Fellow of St. Catherine's College, Oxford University. He received his Ph.D. from Cambridge in 1955. Professor Atiyah was honored with a Fields Medal of the International Congress of Mathematicians in 1966 for his work on the development and application of K-theory, which became a powerful new tool leading to the solution of difficult problems. He has also worked with I.M. Singer in developing the index theory of elliptic differential operators. More recently he has contributed to mathematical aspects of the gauge theories of elementary particle physics. Professor Atiyah is a past president of the London Mathematical Society and the Mathematical Association. He has served as Chairman of the European Mathematical Council since 1978. As a Fellow of the Royal Society, Professor Atiyah was awarded the Copley Medal of the Royal Society in 1988, and has since won numerous other awards, including the De Morgan Medal of the London Mathematical Society, the International Feltrinelli Prize for Mathematical Sciences, and the King Faisal International Prize. Professor Atiyah holds honorary doctorate degrees from the Universities of Bonn, Warwick, Durham, St. Andrew's, Trinity College (Dublin), Chicago, Cambridge, Edinburgh, Essex, London, Sussex, and Ghent. He has served as a visiting lecturer at universities throughout the world, including Harvard, Yale, Columbia, E.T. H. Zurich, Scuola Normale Pisa, the University of Michigan, the University of Illinois, the University of California at Berkeley, Cambridge, Carelton Ottawa, and the French Academy. He was named a Knight Bachelor in 1983.

Special Preface

K-theory is that part of linear algebra that studies additive or abelian properties (e.g., the determinant). Because linear algebra, and its extension to linear analysis, is ubiquitous in mathematics, K-theory has turned out to be useful and relevant in most branches of mathematics. Introduced first in Algebraic Geometry by Grothendieck, it was then transposed to topology by Hirzebruch and myself as expounded in these lecture notes. Subsequently it has wandered into several parts of analysis and, in a different guise, into number theory. It has also been studied systematically in a purely algebraic context. In each of these different fields there are of course different techniques which have to be employed and the relevant K-theory then rests on the appropriate linear algebra or analysis involved.

Most of this development has taken place in the 20 years since this book first appeared and it may be helpful to provide at this stage a brief overview of these new manifestations of K-theory.

The first major application of topological K-theory was to the proof of the index theorem for elliptic differential operators by Singer and myself. K-theory enters into the theorem at two stages. First of all, it enters into the global analysis because of the connection with Fredholm operators explained in the Appendix. Secondly, it enters locally in terms of the symbol of the differential operator. Both of these have their antecedents in Grothendieck's work, but both appearances of K-theory are purely algebraic.

The full development of index theory in its various generalizations makes extensive use of K-theory, and conversely index theory can be used to give proofs of the basic periodicity theorem. In fact, K-theory and index theory really become fused into a single theory in which it is hard to disentangle the topology from the analysis.

In algebraic geometry Grothendieck introduced two K-groups, K^0 and K_0, analogous roughly to cohomology and homology. The K-theory of these notes is the cohomology version, and it seemed clear that it would be useful to find a good definition of the corresponding homology K-theory. I made a tentative start on this programme but the definitive solution was obtained in the work of Brown, Douglas and Fillmore and, independently, by Kasparov. The B-D-F theory started from the study of extensions of C*-algebras by the ideal of compact operators in Hilbert space while the Kasparov theory was even more far-reaching and yielded bivariant functors KK(A,B). These generalize K-groups rather in the way the Grothendieck Ext-groups generalize sheaf cohomology groups.

One feature of these analytical approaches via C*-algebras is that they naturally raise questions about *non-commutative* C*-algebras. In fact, this has proved a profitable line and much work has been done on the non-commutative case. It is rather naturally related to representation theory and this is especially interesting for the C*-algebra of a group.

Further down this line, but in a more differential geometric vein, is the work of A. Connes on foliations. The index theorem for a family of elliptic operators, acting on the fibres of a fibration, has been generalized by Connes to the case of a foliation. There is now no base space (space of leaves) and its role is played by a non-commutative C*-algebra.

An earlier example of this situation arises from infinite covering spaces of compact manifolds. Singer and I had studied the index theorem in such contexts using the K-theory of type II von Neumann algebras studied by M. Breuer. Indices are then real-valued rather than integer-valued and the corresponding K-theory is essentially the usual K-theory tensored with the reals.

Connes' non-commutative differential geometry uses the formulas of cyclic homology and this is related to K-theory in interesting ways that are still being explored.

On the purely topological front K-theory was the first "extraordinary" cohomology theory and it stimulated much work on other such theories. Complex cobordism is the most interesting of these and K-theory is in fact a quotient of it as shown by Conner and Floyd. Moreover both theories fit into the general framework studied by Quillen in which cohomology theories with suitable properties correspond to formal group laws. Cohomology and K-theory correspond to the additive and multiplicative groups respectively, while complex cobordism corresponds to the universal formal group law. Morava has introduced a whole hierarchy of cohomology theories, intermediate between K-theory and complex cobordism. Of these the first corresponds to the formal groups of elliptic curves and has been christened elliptic cohomology. It has many interesting features (studied by Landweber, Stong and Ochanine) and has been related to string theory in physics by Witten. This is currently a very active and exciting field of exploration.

The index theorem for families of Dirac type operators is of great interest in connection with the physics of gauge theories. Motivated partly by this and partly by the ideas of Arakelov and Faltings on "arithmetic surfaces," Soulé and Gillet have been developing a refined K-theory for bundles with connection. This is a hybrid of K-theory and differential geometry and it appears to be a promising avenue for future applications.

A quite different relation between K-theory and number theory was developed some years ago by Quillen. For an arbitrary ring, K^0 and K^1 can be defined using projective modules and automorphisms. Using topological ideas, Quillen showed how to extend these and define "higher K-groups." Unlike the topological K-groups (in which the ring is C(X), the complex-valued continuous functions on a compact space X), the Quillen K-groups are not periodic. Quillen computed his groups for finite fields and there are remarkable conjectures of Lichtenbaum for the K-groups of the ring of integers of a number field. These involve values of L-functions and other number-theoretical quantities. Some parts of the Lichtenbaum conjectures have been established and there is every indication that the number theory aspects of K-theory will be a rich field.

M. F. Atiyah
December, 1988

Contents

Introduction

These notes are based on the course of lectures I gave at Harvard in the fall of 1964. They constitute a self-contained account of vector bundles and K-theory assuming only the rudiments of point-set topology and linear algebra. One of the features of the treatment is that no use is made of ordinary homology or cohomology theory. In fact, rational cohomology is defined in terms of K-theory.

The theory is taken as far as the solution of the Hopf invariant problem and a start is made on the J-homomorphism. In addition to the lecture notes proper, two papers of mine published since 1964 have been reproduced at the end. The first, dealing with operations, is a natural supplement to the material in Chapter III. It provides an alternative approach to operations which is less slick but more fundamental than the Grothendieck method of Chapter III, and it relates operations and filtration. Actually, the lectures deal with compact spaces, not cell-complexes, and so the skeleton-filtration does not figure in the notes. The second paper provides a new approach to real K-theory and so fills an obvious gap in the lecture notes.

CHAPTER I. Vector Bundles

§1.1. Basic definitions. We shall develop the theory of complex vector bundles only, though much of the elementary theory is the same for real and symplectic bundles. Therefore, by vector space, we shall always understand complex vector space unless otherwise specified.

Let X be a topological space. A family of vector spaces over X is a topological space E, together with:

(i) a continuous map $p : E \to X$

(ii) a finite dimensional vector space structure on each

$$E_x = p^{-1}(x) \qquad \text{for } x \in X ,$$

compatible with the topology on E_x induced from E.

The map p is called the projection map, the space E is called the total space of the family, the space X is called the base space of the family, and if $x \in X$, E_x is called the fiber over x.

A section of a family $p : E \to X$ is a continuous map $s : X \to E$ such that $ps(x) = x$ for all $x \in X$.

A homomorphism from one family $p : E \to X$ to another family $q : F \to X$ is a continuous map $\varphi : E \to F$ such that:

(i) $q\varphi = p$

(ii) for each $x \in X$, $\varphi : E_x \to F_x$ is a linear map of vector spaces.

We say that φ is an <u>isomorphism</u> if φ is bijective and φ^{-1} is continuous. If there exists an isomorphism between E and F, we say that they are isomorphic.

<u>Example</u> 1. Let V be a vector space, and let $E = X \times V$, $p : E \to X$ be the projection onto the first factor. E is called the <u>product family</u> with fiber V. If F is any family which is isomorphic to some product family, F is said to be a <u>trivial</u> family.

If Y is a subspace of X, and if E is a family of vector spaces over X with projection p, $p : p^{-1}(Y) \to Y$ is clearly a family over Y. We call it the <u>restriction</u> of E to Y, and denote it by $E|Y$. More generally, if Y is any space, and $f : Y \to X$ is a continuous map, then we define the induced family $f^*(p) : f^*(E) \to Y$ as follows:

$f^*(E)$ is the subspace of $Y \times E$ consisting of all points (y, e) such that $f(y) = p(e)$, together with the obvious projection maps and vector space structures on the fibers. If $g : Z \to Y$, then there is a natural isomorphism $g^*f^*(E) \cong (fg)^*(E)$ given by sending each point of the form (z, e) into the point $(z, g(z), e)$, where $z \in Z$, $e \in E$. If $f : Y \to X$ is an inclusion map, clearly there is an isomorphism $E|Y \cong f^*(E)$ given by sending each $e \in E$ into the corresponding $(p(e), e)$.

A family E of vector spaces over X is said to be locally trivial if every $x \in X$ posesses a neighborhood U such that $E|U$ is trivial. A locally trivial family will also be called a vector bundle. A trivial family will be called a trivial bundle. If $f : Y \to X$, and if E is a vector bundle over X, it is easy to see that $f^*(E)$ is a vector bundle over Y. We shall call $f^*(E)$ the induced bundle in this case.

Example 2. Let V be a vector space, and let X be its associated projective space. We define $E \subset X \times V$ to be the set of all (x, v) such that $x \in X$, $v \in V$, and v lies in the line determining x. We leave it to the reader to show that E is actually a vector bundle.

Notice that if E is a vector bundle over X, then $\dim(E_x)$ is a locally constant function on X, and hence is a constant on each connected component of X. If $\dim(E_x)$ is a constant on the whole of X, then E is said to have a dimension, and the dimension of E is the common number $\dim(E_x)$ for all x. (Caution: the dimension of E so defined is usually different from the dimension of E as a topological space.)

Since a vector bundle is locally trivial, any section of a vector bundle is locally described by a vector valued function on the base space. If E is a vector bundle, we denote by $\Gamma(E)$ the set of all sections of E. Since the set of functions on a space with values in a fixed vector space is itself a vector space, we see that $\Gamma(E)$ is a vector space in a natural way.

Suppose that V, W are vector spaces, and that $E = X \times V$, $F = X \times W$ are the corresponding product bundles. Then any homomorphism $\varphi : E \to F$ determines a map $\Phi : X \to \mathrm{Hom}(V, W)$ by the formula $\varphi(x, v) = (x, \Phi(x)v)$. Moreover, if we give Hom(V, W) its usual topology, then Φ is continuous; conversely, any such continuous map $\Phi : X \to \mathrm{Hom}(V, W)$ determines a homomorphism $\varphi : E \to F$. (This is most easily seen by taking bases $\{e_i\}$ and $\{f_i\}$ for V and W respectively. Then each $\Phi(x)$ is represented by a matrix $\Phi(x)_{i,j}$, where

$$\Phi(x)e_i = \sum_j \Phi(x)_{i,j} f_j \quad .$$

The continuity of either φ or Φ is equivalent to the continuity of the functions $\Phi_{i,j}$.)

Let $\mathrm{Iso}(V, W) \subset \mathrm{Hom}(V, W)$ be the subspace of all isomorphisms between V and W. Clearly, Iso(V, W) is an open set in Hom(V, W). Further, the inverse map $T \to T^{-1}$ gives us a continuous map $\mathrm{Iso}(V, W) \to \mathrm{Iso}(W, V)$. Suppose that $\varphi : E \to F$ is such that $\varphi_x : E_x \to F_x$ is an isomorphism for all $x \in X$. This is equivalent to the statement that $\Phi(X) \subset \mathrm{Iso}(V, W)$. The map $x \to \Phi(x)^{-1}$ defines $\Psi : X \to \mathrm{Iso}(W, V)$, which is continuous. Thus the corresponding map $\psi : F \to E$ is continuous. Thus

$\varphi : E \to F$ is an isomorphism if and only if it is bijective or, equivalently, φ is an isomorphism if and only if each φ_x is an isomorphism. Further, since $\mathrm{Iso}(V,W)$ is open in $\mathrm{Hom}(V,W)$, we see that for any homomorphism φ, the set of those points $x \in X$ for which φ_x is an isomorphism form an open subset of X. All of these assertions are local in nature, and therefore are valid for vector bundles as well as for trivial families.

Remark: The finite dimensionality of V is basic to the previous argument. If one wants to consider infinite dimensional vector bundles, then one must distinguish between the different operator topologies on $\mathrm{Hom}(V,W)$.

§1.2. Operations on vector bundles. Natural operations on vector spaces, such as direct sum and tensor product, can be extended to vector bundles. The only troublesome question is how one should topologize the resulting spaces. We shall give a general method for extending operations from vector spaces to vector bundles which will handle all of these problems uniformly.

Let T be a functor which carries finite dimensional vector spaces into finite dimensional vector spaces. For simplicity, we assume that T is a covariant functor of one variable. Thus, to every vector space V, we have an associated vector space T(V). We shall say that T is a continuous functor if for all V and W, the map $T : \mathrm{Hom}(V,W) \to \mathrm{Hom}(T(V),T(W))$ is continuous.

If E is a vector bundle, we define the set T(E) to be the union

$$\bigcup_{x \in X} T(E_x) ,$$

and, if $\varphi : E \to F$, we define $T(\varphi) : T(E) \to T(F)$ by the maps $T(\varphi_x) : T(E_x) \to T(F_x)$. What we must show is that T(E) has a natural topology, and that, in this topology, $T(\varphi)$ is continuous.

We begin by defining T(E) in the case that E is a product bundle. If $E = X \times V$, we define T(E) to be $X \times T(V)$ in the product topology. Suppose that $F = X \times W$, and that $\varphi : E \to F$ is a homomorphism. Let $\Phi : X \to \mathrm{Hom}(V, W)$ be the corresponding map. Since, by hypothesis, $T : \mathrm{Hom}(V, W) \to \mathrm{Hom}(T(V), T(W))$ is continuous, $T\Phi : X \to \mathrm{Hom}(T(V), T(W))$ is continuous. Thus $T(\varphi): X \times T(V) \to X \times T(W)$ is also continuous. If φ is an isomorphism, then $T\varphi$ will be an isomorphism since it is continuous and an isomorphism on each fiber.

Now suppose that E is trivial, but has no preferred product structure. Choose an isomorphism $\alpha : E \to X \times V$, and topologize $T(E)$ by requiring $T(\alpha) : T(E) \to X \times T(V)$ to be a homeomorphism. If $\beta : E \to X \times W$ is any other isomorphism, by letting $\varphi = \beta\alpha^{-1}$ above, we see that $T(\alpha)$ and $T(\beta)$ induce the same topology on $T(E)$, since $T(\varphi) = T(\beta)T(\alpha)^{-1}$ is a homeomorphism. Thus, the topology on E does not depend on the choice of α. Further, if $Y \subset X$, it is clear that the topology on $T(E)|Y$ is the same as that on $T(E|Y)$. Finally, if $\varphi : E \to F$ is a homomorphism of trivial bundles, we see that $T(\varphi) : T(E) \to T(F)$ is continuous, and therefore is a homomorphism.

Now suppose that E is any vector bundle. Then if $U \subset X$ is such that $E|U$ is trivial, we topologize $T(E|U)$ as above. We topologize $T(E)$ by taking for the open sets, those subsets $V \subset T(E)$ such that $V \cap (T(E)|U)$ is open in $T(E|U)$ for all open $U \subset X$ for which $E|U$ is trivial. The reader can now easily verify that if $Y \subset X$, the topology on $T(E|Y)$ is the same as that on $T(E)|Y$, and that, if $\varphi : E \to F$ is any homomorphism, $T(\varphi) : T(E) \to T(F)$ is also a homomorphism.

If $f : Y \to X$ is a continuous map and E is a vector bundle over X then, for any continuous functor T, we have a natural isomorphism

$$f^* T(E) \cong T f^*(E) \quad .$$

The case when T has several variables both covariant and contravariant, proceeds similarly. Therefore we can define for vector bundles E, F corresponding bundles:

(i) $E \oplus F$, their direct sum

(ii) $E \otimes F$, their tensor product

(iii) $\mathrm{Hom}(E, F)$

(iv) E^*, the dual bundle of E

(v) $\lambda^i(E)$, where λ^i is the ith exterior power.

We also obtain natural isomorphisms

(i) $E \oplus F \cong F \oplus E$

(ii) $E \otimes F \cong F \otimes E$

(iii) $E \otimes (F' \oplus F'') \cong (E \otimes F') \oplus (E \otimes F'')$

(iv) $\mathrm{Hom}(E, F) \cong E^* \otimes F$

(v) $\lambda^k(E \oplus F) \cong \bigoplus_{i+j=k} (\lambda^i(E) \otimes \lambda^j(F))$.

Finally, notice that sections of Hom(E, F) correspond in a 1 - 1 fashion with homomorphisms $\varphi : E \to F$. We therefore define HOM(E,F) to be the vector space of all homomorphisms from E to F, and make the identification $HOM(E,F) = \Gamma(Hom(E,F))$.

§ 1.3. Sub-bundles and quotient bundles. Let E be a vector bundle. A sub-bundle of E is a subset of E which is a bundle in the induced structure.

A homomorphism $\varphi : F \to E$ is called a monomorphism (respectively epimorphism) if each $\varphi_x : F_x \to E_x$ is a monomorphism (respectively epimorphism). Notice that $\varphi : F \to E$ is a monomorphism if and only if $\varphi^* : E^* \to F^*$ is an epimorphism. If F is a sub-bundle of E , and if $\varphi : F \to E$ is the inclusion map, then φ is a monomorphism.

LEMMA 1.3.1. If $\varphi : F \to E$ is a monomorphism, then $\varphi(F)$ is a sub-bundle of E , and $\varphi : F \to \varphi(F)$ is an isomorphism.

Proof: $\varphi : F \to \varphi(F)$ is a bijection, so if $\varphi(F)$ is a sub-bundle, φ is an isomorphism. Thus we need only show that $\varphi(F)$ is a sub-bundle.

The problem is local, so it suffices to consider the case when E and F are product bundles . Let $E = X \times V$ and let $x \in X$; choose $W_x \subset V$ to be a subspace complementary to $\varphi(F_x)$. $G = X \times W_x$ is a sub-bundle of E . Define $\theta : F \oplus G \to E$ by $\theta(a \oplus b) = \varphi(a) + i(b)$, where $i : G \to E$ is the inclusion. By construction, θ_x is an isomorphism. Thus, there exists an open neighborhood U of x such that $\theta | U$ is an isomorphism. F is a sub-bundle of $F \oplus G$, so $\theta(F) = \varphi(F)$ is a sub-bundle of $\theta(F \oplus G) = E$ on U .

Notice that in our argument, we have shown more than we have stated. We have shown that if $\varphi : F \to E$, then the set of points for which φ_x is a monomorphism form an open set. Also, we have shown that, locally, a sub-bundle is a direct summand. This second fact allows us to define quotient bundles.

DEFINITION 1.3.1. If F is a sub-bundle of E, the quotient bundle E/F is the union of all the vector spaces E_x/F_x, given the quotient topology.

Since F is locally a direct summand in E, we see that E/F is locally trivial, and thus is a bundle. This justifies the terminology.

If $\varphi : F \to E$ is an arbitrary homomorphism, the function dimension(kernel (φ_x)) need not be constant, or even locally constant.

DEFINITION 1.3.2. $\varphi : F \to E$ is said to be a <u>strict</u> homomorphism if dimension(kernel(φ_x)) is locally constant.

PROPOSITION 1.3.2. <u>If</u> $\varphi : F \to E$ <u>is strict, then:</u>

(i) $\text{kernel}(\varphi) = \bigcup_x \text{kernel}(\varphi_x)$ <u>is a sub-bundle of</u> F

(ii) $\text{image}(\varphi) = \bigcup_x \text{image}(\varphi_x)$ <u>is a sub-bundle of</u> E

(iii) $\text{cokernel}(\varphi) = \bigcup_x \text{cokernel}(\varphi_x)$ <u>is a bundle in the quotient structure.</u>

<u>Proof:</u> Notice that (ii) implies (iii). We first prove (ii). The problem is local, so we can assume $F = X \times V$ for some V. Given $x \in X$, we choose $W_x \subset V$ complementary to $\ker(\varphi_x)$ in V. Put $G = X \times W_x$; then φ induces, by composition with the inclusion, a homomorphism $\psi : G \to E$, such that ψ_x is a monomorphism. Thus, ψ is a monomorphism in some neighborhood U of x. Therefore, $\psi(G)|U$ is a sub-bundle of $E|U$. However, $\psi(G) \subset \varphi(F)$, and since $\dim(\varphi(F_y))$ is constant for all y, and $\dim(\psi(G_y)) = \dim(\psi(G_x)) = \dim(\varphi(F_x)) = \dim(\varphi(F_y))$ for all $y \in U$, $\psi(G)|U = \varphi(F)|U$. Thus $\varphi(F)$ is a sub-bundle of E.

Finally, we must prove (i). Clearly, $\varphi^* : E^* \to F^*$ is strict. Since $F^* \to \operatorname{coker}(\varphi^*)$ is an epimorphism, $(\operatorname{coker}(\varphi^*))^* \to F^{**}$ is a monomorphism. However, for each x we have a natural commutative diagram

$$\begin{array}{ccc} \ker(\varphi_x) & \longrightarrow & F_x \\ \downarrow & & \downarrow \\ (\operatorname{coker} \varphi_x^*)^* & \longrightarrow & F_x^{**} \end{array}$$

in which the vertical arrows are isomorphisms. Thus $\ker(\varphi) \cong (\operatorname{coker}(\varphi^*))^*$ and so, by (1.3.1), is a sub-bundle of F.

Again, we have proved something more than we have stated. Our argument shows that for any $x \in X$, $\dim \varphi_x(F_x) \leq \dim \varphi_y(F_y)$

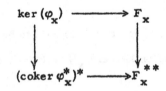

for all $y \in U$, U some neighborhood of x. Thus, rank (φ_x) is an upper semi-continuous function of x.

DEFINITION 1.3.3. A projection operator $P : E \to E$ is a homomorphism such that $P^2 = P$.

Notice that $\text{rank}(P_x) + \text{rank}(1 - P_x) = \dim E_x$ so that, since both $\text{rank}(P_x)$ and $\text{rank}(1 - P_x)$ are upper semi-continuous functions of x, they are locally constant. Thus both P and $1 - P$ are strict homomorphisms. Since $\ker(P) = (1 - P)E$, E is the direct sum of the two sub-bundles PE and $(1 - P)E$. Thus any projection operator $P : E \to E$ determines a direct sum decomposition $E = (PE) \oplus ((1 - P)E)$.

We now consider metrics on vector bundles. We define a functor Herm which assigns to each vector space V the vector space $\text{Herm}(V)$ of all Hermitian forms on V. By the techniques of §1.2, this allows us to define a vector bundle $\text{Herm}(E)$ for every bundle E.

DEFINITION 1.3.4. A metric on a bundle E is any section $h : X \to \text{Herm}(E)$ such that $h(x)$ is positive definite for all $x \in X$. A bundle with a *specified metric* is called a Hermitian bundle.

Suppose that E is a bundle, F is a sub-bundle of E, and that h is a Hermitian metric on E. Then for each $x \in X$

we consider the orthogonal projection $P_x : E_x \to F_x$ defined by the metric. This defines a map $P : E \to F$ which we shall now check is continuous. The problem being local we may assume F is trivial, so that we have sections $f_1, \cdots, f_n$ of F giving a basis in each fiber. Then for $v \in F_x$ we have

$$P_x(v) = \sum_i h_x(v, f_i(x))f_i(x) \quad .$$

Since h is continuous this implies that P is continuous. Thus P is a projection operator on E. If $F_x^{\perp}$ is the subspace of E_x which is orthogonal to F_x under h, we see that $F^{\perp} = \bigcup_x F_x^{\perp}$ is the kernel of P, and thus is a sub-bundle of E, and that $E \cong F \oplus F^{\perp}$. Thus, a metric provides any sub-bundle with a definite complementary sub-bundle.

Remark: So far, most of our arguments have been of a very general nature, and we could have replaced "continuous" with "algebraic", "differentiable", "analytic", etc. without any trouble. In the next section, our arguments become less general.

§ 1.4. Vector bundles on compact spaces. In order to proceed further, we must make some restriction on the sort of base spaces which we consider. We shall assume from now on that our base spaces are compact Hausdorff. We leave it to the reader to notice which results hold for more general base spaces.

Recall that if $f : X \to V$ is a continuous vector-valued function, the support of f (written supp. f) is the closure of $f^{-1}(V - \{0\})$.

We need the following results from point set topology. We state them in vector forms which are clearly equivalent to the usual forms.

Tietze Extension Theorem. Let X be a normal space, $Y \subset X$ a closed subspace, V a real vector space, and $f : Y \to V$ a continuous map. Then there exists a continuous map $g : X \to V$ such that $g|Y = f$.

Existence of Partitions of Unity. Let X be a compact Hausdorff space, $\{U_i\}$ a finite open covering. Then there exist continuous maps $f_i : X \to R$ such that:

(i) $f_i(x) \geq 0$ all $x \in X$

(ii) supp $(f_i) \subset U_i$

(iii) $\sum_i f_i(x) = 1$ all $x \in X$.

Such a collection $\{f_i\}$ is called a partition of unity.

We first give a bundle form of the Tietze extension theorem.

LEMMA 1.4.1. Let X be compact Hausdorff, $Y \subset X$ a closed subspace, and E a bundle over X. Then any section $s : Y \to E|Y$ can be extended to X.

Proof: Let $s \in \Gamma(E|Y)$. Since, locally, s is a vector-valued function, we can apply the Tietze extension theorem to show that for each $x \in X$, there exists an open set U containing x and $t \in \Gamma(E|U)$ such that $t|U \cap Y = s|U \cap Y$. Since X is compact, we can find a finite subcover $\{U_\alpha\}$ by such open sets. Let $t_\alpha \in \Gamma(E|U_\alpha)$ be the corresponding sections and let $\{p_\alpha\}$ be a partition of unity with $\operatorname{supp}(p_\alpha) \subset U_\alpha$. We define $S_\alpha \in \Gamma(E)$ by

$$
\begin{aligned}
S_\alpha(x) &= p_\alpha(x)\, t_\alpha(x) && \text{if } x \in U_\alpha \\
&= 0 && \text{otherwise}.
\end{aligned}
$$

Then ΣS_α is a section of E and its restriction to Y is clearly s.

LEMMA 1.4.2. Let Y be a closed subspace of a compact Hausdorff space X, and let E, F be two vector bundles over X. If $f : E|Y \longrightarrow F|Y$ is an isomorphism, then there exists an opet set U containing Y and an extension $f : E|U \longrightarrow F|U$ which is an isomorphi

Proof: f is a section of $\mathrm{Hom}(E|Y, F|Y)$, and thus, extends to a section of $\mathrm{Hom}(E, F)$. Let U be the set of those points for which this map is an isomorphism. Then U is open and contains Y.

LEMMA 1.4.3. *Let Y be a compact Hausdorff space, $f_t : Y \to X$ $(0 \leq t \leq 1)$ a homotopy and E a vector bundle over X. Then*

$$f_0^* E \cong f_1^* E .$$

Proof: If I denotes the unit interval let $f : Y \times I \to X$ be the homotopy, so that $f(y, t) = f_t(y)$, and let $\pi : Y \times I \to Y$ denote the projection. Now apply (1.4.2) to the bundles f^*E, $\pi^* f_t^* E$ and the subspace $Y \times \{t\}$ of $Y \times I$, on which there is an obvious isomorphism s. By the compactness of Y we deduce that f^*E and $\pi^* f_t^* E$ are isomorphic in some strip $Y \times \delta t$ where δt denotes a neighborhood of $\{t\}$ in I. Hence the isomorphism class of $f_t^* E$ is a locally constant function of t. Since I is connected this implies it is constant, whence

$$f_0^* E \cong f_1^* E .$$

We shall use $\mathrm{Vect}(X)$ to denote the set of isomorphism classes of vector bundles on X, and $\mathrm{Vect}_n(X)$ to denote the subset of $\mathrm{Vect}(X)$ given by bundles of dimension n. $\mathrm{Vect}(X)$ is an abelian semi-group under the operation $\oplus$. In $\mathrm{Vect}_n(X)$ we have one naturally distinguished element - the class of the trivial bundle of dimension n .

LEMMA 1.4.4.

(1) If $f : X \to Y$ is a homotopy equivalence, $f^* : \mathrm{Vect}(Y) \to \mathrm{Vect}(X)$ is bijective .

(2) If X is contractible, every bundle over X is trivial and $\mathrm{Vect}(X)$ is isomorphic to the non-negative integers .

LEMMA 1.4.5. If E is a bundle over $X \times I$, and $\pi : X \times I \to X \times \{0\}$ is the projection, E is isomorphic to $\pi^*(E|X \times \{0$

Both of these lemmas are immediate consequences of (1.4.3) .

Suppose now Y is closed in X , E is a vector bundle over X and $\alpha : E|Y \to Y \times V$ is an isomorphism. We refer to α as a trivialization of E over Y . Let $\pi : Y \times V \to V$ denote the projection and define an equivalence relation on $E|Y$ by

$$e \sim e' \Longleftrightarrow \pi\alpha(e) = \pi\alpha(e') \quad .$$

We extend this by the identity on $E|X - Y$ and we let E/α denote the quotient space of E given by this equivalence relation. It has a natural structure of a family of vector spaces over X/Y . We assert that E/α is in fact a vector bundle. To see this we have only to verify the local trivality at the base point Y/Y of X/Y. Now by (1.4.2) we can extend α to an isomorphism $\tilde{\alpha} : E|U \to U \times V$ for some open set U containing Y. Then $\tilde{\alpha}$ induces an isomorphism

$$(E|U)/\alpha \cong (U/Y) \times V$$

which establishes the local triviality of E/α.

Suppose α_0, α_1 are homotopic trivializations of E over Y. This means that we have a trivialization β of $E \times I$ over $Y \times I \subset X \times I$ inducing α_0 and α_1 at the two end points of I. Let $f: (X/Y) \times I \to (X \times I)/(Y \times I)$ be the natural map. Then $f^*(E \times I/\beta)$ is a bundle on $(X/Y) \times I$ whose restriction to $(X/Y) \times \{i\}$ is E/α_i $(i = 0, 1)$. Hence, by (1.4.3),

$$E/\alpha_0 \cong E/\alpha_1 \quad .$$

To summarize we have established

LEMMA 1.4.7. <u>A trivialization</u> α <u>of a bundle</u> E <u>over</u> $Y \subset X$ <u>defines a bundle</u> E/α <u>over</u> X/Y. <u>The isomorphism class of</u> E/α <u>depends only on the homotopy class of</u> α.

Using this we shall now prove

LEMMA 1.4.8. <u>Let</u> $Y \subset X$ <u>be a closed contractible subspace. Then</u> $f : X \to X/Y$ <u>induces a bijection</u> $f^* : \mathrm{Vect}(X/Y) \to \mathrm{Vect}(X)$.

<u>Proof:</u> Let E be a bundle on X then by (1.4.4) $E|Y$ is trivial. Thus trivializations $\alpha : E|Y \to Y \times V$ exist. Moreover, two such trivializations differ by an automorphism of $Y \times V$, i.e., by a map $Y \to GL(V)$. But $GL(V) = GL(n, C)$ is connected and V is contractible. Thus α is unique up to homotopy and so the isomorphism class of $E|\alpha$ is uniquely determined by that of E. Thus we have constructed a map

$$\text{Vect}(X) \longrightarrow \text{Vect}(X/Y)$$

and this is clearly a two-sided inverse for f^*. Hence f^* is bijective as asserted.

Vector bundles are frequently constructed by a glueing or clutching construction which we shall now describe. Let

$$X = X_1 \cup X_2, \qquad A = X_1 \cap X_2,$$

all the spaces being compact. Assume that E_i is a vector bundle over X_i and that $\varphi : E_1|A \to E_2|A$ is an isomorphism. Then we define the vector bundle $E_1 \cup_\varphi E_2$ on X as follows. As a topological space $E_1 \cup_\varphi E_2$ is the quotient of the disjoint sum $E_1 + E_2$ by the equivalence relation which identifies $e_1 \in E_1|A$ with $\varphi(e_1) \in E_2|A$. Identifying X with the corresponding quotient of $X_1 + X_2$ we obtain a natural projection $p : E_1 \cup_\varphi E_2 \to X$, and $p^{-1}(x)$ has a natural vector space structure. It remains to show that $E_1 \cup_\varphi E_2$ is locally trivial. Since $E_1 \cup_\varphi E_2 | X - A = (E_1|X_1 - A) + (E_2|X_2 - A)$ the local triviality at points $x \notin A$ follows from that of E_1 and E_2. Therefore, let $a \in A$ and let V_1 be a closed neighborhood of a in X_1 over which E_1 is trivial, so that we have an isomorphism

$$\theta_1 : E_1|V_1 \to V_1 \times \mathbb{C}^n \quad .$$

Restricting to A we get an isomorphism

$$\theta_1^A : E_1|V_1 \cap A \to (V_1 \cap A) \times \mathbb{C}^n \quad .$$

Let
$$\theta_2^A : E_2|V_1 \cap A \to (V_1 \cap A) \times \mathbb{C}^n$$

be the isomorphism corresponding to θ_1^A under φ. By (1.4.2) this can be extended to an isomorphism

$$\theta_2 : E_2|V_2 \to V_2 \times \mathbb{C}^n$$

where V_2 is a neighborhood of a in X_2. The pair θ_1, θ_2 then defines in an obvious way an isomorphism

$$\theta_1 \cup_\varphi \theta_2 : E_1 \cup_\varphi E_2 | V_1 \cup V_2 \longrightarrow (V_1 \cup V_2) \times \mathbb{C}^n \; ,$$

establishing the local triviality of $E_1 \cup_\varphi E_2$.

Elementary properties of this construction are the following:

(i) If E is a bundle over X and $E_i = E|X_i$, then the identity defines an isomorphism $I_A : E_1|A \to E_2|A$, and

$$E_1 \cup_{I_A} E_2 \cong E .$$

(ii) If $\beta_i : E_i \to E_i'$ are isomorphisms on X_i and $\varphi' \beta_1 = \beta_2 \varphi$, then

$$E_1 \cup_\varphi E_2 \cong E_1' \cup_{\varphi'} E_2' .$$

(iii) If (E_i, φ) and (E_i', φ') are two "clutching data" on the X_i, then

$$(E_1 \cup_\varphi E_2) \oplus (E_1' \cup_{\varphi'} E_2') \cong E_1 \oplus E_1' \underset{\varphi \oplus \varphi'}{\cup} E_2 \oplus E_2' ,$$

$$(E_1 \cup_\varphi E_2) \otimes (E_1' \cup_{\varphi'} E_2') \cong E_1 \otimes E_1' \underset{\varphi \otimes \varphi'}{\cup} E_2 \otimes E_2' ,$$

$$(E_1 \cup_\varphi E_2)^* \cong E_1^* \cup_{(\varphi^*)^{-1}} E_2^* .$$

Moreover, we also have

LEMMA 1.4.6. <u>The isomorphism class of $E_1 \cup_\varphi E_2$ depends only on the homotopy class of the isomorphism</u> $\varphi : E_1|A \to E_2|A$.

<u>Proof</u>: A homotopy of isomorphisms $E_1|A \to E_2|A$ means an isomorphism

$$\Phi : \pi^* E_1|A \times I \to \pi^* E_2|A \times I ,$$

where I is the unit interval and $\pi : X \times I \to X$ is the projection. Let

$$f_t : X \longrightarrow X \times I$$

be defined by $f_t(x) = x \times \{t\}$ and denote by

$$\varphi_t : E_1|A \to E_2|A$$

the isomorphism induced from Φ by f_t. Then

$$E_1 \cup_{\varphi_t} E_2 \cong f_t^*(\pi^* E_1 \cup_\Phi \pi^* E_2) .$$

Since f_0 and f_1 are homotopic it follows from (1.4.3) that

$$E_1 \cup_{\varphi_0} E_2 \cong E_1 \cup_{\varphi_1} E_2$$

as required.

Remark: The "collapsing" and "clutching" constructions for bundles (on X/Y and $X_1 \cup X_2$ respectively) are both special cases of a general process of forming bundles over quotient spaces. We leave it as an exercise to the reader to give a precise general formulation.

We shall denote by $[X, Y]$ the set of homotopy classes of maps $X \to Y$.

LEMMA 1.4.9. For any X, there is a natural isomorphism $\mathrm{Vect}_n(S(X)) \cong [X, GL(n,C)]$.

Proof: Write $S(X)$ as $C^+(X) \cup C^-(X)$, where $C^+(X) = [0, 1/2] \times X/\{0\} \times X$, $C^-(X) = [1/2, 1] \times X/\{1\} \times X$. Then $C^+(X) \cap C^-(X) = X$. If E is any n-dimensional bundle over $S(X)$, $E|C^+(X)$ and $E|C^-(X)$ are trivial. Let $\alpha^{\pm} : E|C^{\pm}(X) \cong C^{\pm}(X) \times V$ be such isomorphisms. Then $(\alpha^+|X)(\alpha^-|X)^{-1} : X \times V \to X \times V$ is a bundle map, and thus defines a map α of X into $GL(n,\mathbb{C}) = \mathrm{Iso}(V)$. Since both $C^+(X)$ and $C^-(X)$ are contractible, the homotopy classes of both α^+ and α^- are well defined, and thus the homotopy class of α is well defined. Thus we have a natural map $\theta : \mathrm{Vect}_n(S(X)) \to [X, GL(n, \mathbb{C})]$. The clutching construction on the other hand defines by (1.4.6) a map

$$\varphi : [X, GL(n, C)] \longrightarrow \mathrm{Vect}_n(S(X)) .$$

It is clear that θ and φ are inverses of each other and so are bijections.

We have just seen that $\mathrm{Vect}_n(S(X))$ has a homotopy theoretic interpretation. We now give a similar interpretation to $\mathrm{Vect}_n(X)$. First we must establish some simple facts about quotient bundles.

LEMMA 1.4.10. Let E be any bundle over X. Then there exists a (Hermitian) metric on E.

__Proof:__ A metric on a vector space V defines a metric on the product bundle $X \times V$. Hence metrics exist on trivial bundles. Let $\{U_\alpha\}$ be a finite open covering of X such that $E|U_\alpha$ is trivial and let h_α be a metric for $E|U_\alpha$. Let $\{p_\alpha\}$ be a partition of unity with supp. $p_\alpha \subset U_\alpha$ and define

$$\begin{aligned} k_\alpha(x) &= p_\alpha(x)\, h_\alpha(x) && \text{for } x \in U_\alpha \\ &= 0 && \text{otherwise.} \end{aligned}$$

Then k_α is a section of Herm(E) and is positive semi-definite. But for any $x \in X$ there exists α such that $p_\alpha(x) > 0$ (since $\Sigma p_\alpha = 1$) and so $x \in U_\alpha$. Hence, for this α, $k_\alpha(x)$ is positive definite. Hence $\Sigma_\alpha k_\alpha(x)$ is positive definite for all $x \in X$ and so $k = \Sigma k_\alpha$ is a metric for E.

A sequence of vector bundle homomorphisms

$$\longrightarrow E \longrightarrow F \longrightarrow \cdots$$

is called __exact__ if for each $x \in X$ the sequence of vector space homomorphisms

$$\longrightarrow E_x \longrightarrow F_x \longrightarrow \cdots$$

is exact.

COROLLARY 1.4.11. __Suppose that__ $0 \longrightarrow E' \xrightarrow{\varphi'} E \xrightarrow{\varphi''} E'' \longrightarrow 0$ __is an exact sequence of bundles over__ X. __Then there exists an isomorphism__ $E \cong E' \oplus E''$.

Proof: Give E a metric. Then $E \cong E' \oplus (E')^{\perp}$. However, $(E')^{\perp} \cong E''$.

A subspace $V \subset \Gamma(E)$ is said to be ample if

$$\varphi : X \times V \longrightarrow E$$

is a surjection, where $\varphi(x, s) = s(x)$.

LEMMA 1.4.12. *If E is any bundle over a compact Hausdorff space X, then $\Gamma(E)$ contains a finite dimensional ample subspace.*

Proof: Let $\{U_\alpha\}$ be a finite open covering of X so that $E|U_\alpha$ is trivial for each α, and let $\{p_\alpha\}$ be a partition of unity with supp $p_\alpha \subset U_\alpha$. Since $E|U_\alpha$ is trivial we can find a finite-dimensional ample subspace $V_\alpha \subset \Gamma(E|U_\alpha)$. Now define

$$\theta_\alpha : V_\alpha \longrightarrow \Gamma(E)$$

by

$$\begin{aligned} \theta_\alpha v_\alpha(x) &= p_\alpha(x) \cdot v_\alpha(x) && \text{if } x \in U_\alpha \\ &= 0 && \text{otherwise}. \end{aligned}$$

The θ_α define a homomorphism

$$\theta : \prod_\alpha V_\alpha \longrightarrow \Gamma(E)$$

and the image of θ is a finite dimensional subspace of $\Gamma(E)$; in fact, for each $x \in X$ there exists α with $p_\alpha(x) > 0$ and

and so the map

$$\theta_\alpha(V_\alpha) \longrightarrow E_x$$

is surjective.

COROLLARY 1.4.13. If E is any bundle, there exists an epimorphism $\varphi : X \times C^m \longrightarrow E$ for some integer m.

COROLLARY 1.4.14. If E is any bundle, there exists a bundle F such that $E \oplus F$ is trivial.

We are now in a position to prove the existence of a homotopy theoretic definition for $\mathrm{Vect}_n(X)$. We first introduce Grassmann manifolds. If V is any vector space, and n any integer, the set $G_n(V)$ is the set of all subspaces of V of codimension n. If V is given some Hermitian metric, each subspace of V determines a projection operator. This defines a map $G_n(V) \to \mathrm{End}(V)$, where $\mathrm{End}(V)$ is the set of endomorphisms of V. We give $G_n(V)$ the topology induced by this map.

Suppose that E is a bundle over a space X, V is a vector space, and $\varphi : X \times V \to E$ is an epimorphism. If we map X into $G_n(V)$ by assigining to x the subspace $\ker(\varphi_x)$, this map is continuous for any metric on V (here $n = \dim(E)$). We call the map $X \to G_n(V)$ the map induced by φ.

Let V be a vector space, and let $F \subset G_n(V) \times V$ be the sub-bundle consisting of all points (g, v) such that $v \in g$. Then, if $E = (G_n(V) \times V)/F$ is the quotient bundle, E is called the classifying bundle over $G_n(V)$.

Notice that if E' is a bundle over X, and $\varphi : X \times V \to E'$ is an epimorphism, then if $f : X \to G_n(V)$ is the map induced by φ, we have $E' \cong f^*(E)$, where E is the classifying bundle.

Suppose that h is a metric on V. We denote by $G_n(V_h)$ the set $G_n(V)$ with the topology induced by h. If h' is another metric on V, then the epimorphism $G_n(V_h) \times V \to E$ (where E is the classifying bundle) induces the identity map $G_n(V_h) \to G_n(V_{h'})$. Thus the identity map is continuous. Thus, the topology on $G_n(V)$ does not depend on the metric.

Now consider the natural projections

$$C^m \longrightarrow C^{m-1}$$

given by $(z_1, \cdots, z_m) \to (z_1, \cdots, z_{m-1})$. These induce continuous maps

$$\iota_{m-1} : G_n(C^{m-1}) \longrightarrow G_n(C^m) \quad .$$

If $E_{(m)}$ denotes the classifying bundle over $G_n(C^m)$ it is immediate that

$$\iota^*_{m-1}(E_m) \cong E_{(m-1)} \quad .$$

THEOREM 1.4.15. *The map*

$$\xrightarrow[m]{\lim} [X, G_n(C^m)] \longrightarrow \mathrm{Vect}_n(X)$$

induced by $f \to f^*(E_{(m)})$ *for* $f : X \to G_n(C^m)$, *is an isomorphism for all compact Hausdorff spaces* X.

Proof: We shall construct an inverse map. If E is a bundle over X, there exists (by (1.4.13)) an epimorphism $\varphi : X \times C^m \to E$. Let $f : X \to G_n(C^m)$ be the map induced by φ. If we can show that the homotopy class of f (in $G_n(V^{m'})$ for m' sufficiently large does not depend on the choice of φ, then we construct our inverse map $\mathrm{Vect}_n(X) \to \xrightarrow[m]{\lim} [X, G_n(V^m)]$ by sending E to the homotopy class of f.

Suppose that $\varphi_i : X \times C^{m_i} \to E$ are two epimorphisms $(i = 0, 1)$. Let $g_i : X \to G_n(C^{m_i})$ be the map induced by φ_i. Define $\psi_t : X \times C^{m_0} \times C^{m_1} \to E$ by $\psi_t(x, v_0, v_1) = (1 - t)\,\varphi_0(x, v_0) + t\varphi_1(x, v_1)$. This is an epimorphism. Let $f_t : X \to G_n(C^{m_0} \oplus C^{m_1})$ be the map induced by ψ_t. If we identify $C^{m_0} \oplus C^{m_1}$ with $C^{m_0+m_1}$ by $(z_1, \cdots, z_{m_0}) \oplus (u_1, \cdots, u_{m_1}) \longrightarrow (z_1, \cdots z_{m_0}, \cdots, u_{m_1})$ then

$$f_0 = j_0 g_0 , \qquad f_1 = T\, j_1 g_1 ,$$

where $j_i : G_n(C^{m_i}) \to G_n(C^{m_0+m_1})$ is the natural inclusion and

$$T : G_n(C^{m_0+m_1}) \longrightarrow G_n(C^{m_0+m_1})$$

is the map induced by a permutation of coordinates in $C^{m_0+m_1}$, and so is homotopic to the identity. Hence $j_1 g_1$ is homotopic to f_1 and hence to $j_0 g_0$ as required.

Remark. It is possible to interpret vector bundles as modules in the following way. Let $C(X)$ denote the ring of continuous complex-valued functions on X. If E is a vector bundle over X then $\Gamma(E)$ is a $C(X)$ - module under point-wise multiplication, i. e.,

$$fs(x) = f(x)s(x) \qquad f \in C(X), \quad s \in \Gamma(E) .$$

Moreover a homomorphism $\varphi : E \to F$ determines a $C(X)$-module homomorphism

$$\Gamma\varphi : \Gamma(E) \longrightarrow \Gamma(F) .$$

Thus Γ is a functor from the category $\mathcal{V}$ of vector bundles over X to the category $\mathfrak{M}$ of $C(X)$-modules. If E is trivial of dimension n, then $\Gamma(E)$ is free of rank n. If F is also trivial then

$$\Gamma : \mathrm{HOM}(E, F) \longrightarrow \mathrm{Hom}_{C(X)}(\Gamma(E), \Gamma(F))$$

is bijective. In fact, choosing isomorphisms $E \cong X \times V$,

$F \cong X \times W$ we have

$$\mathrm{HOM}(E, F) \cong \mathrm{Hom}_C(V, W)^X \cong C(X) \otimes \mathrm{Hom}_C(V, W)$$
$$\cong \mathrm{Hom}_{C(X)}(\Gamma(E), \Gamma(F)) \quad .$$

Thus Γ induces an equivalence between the category $\mathfrak{T}$ of trivial vector bundles to the category $\mathfrak{F}$ of free $C(X)$-modules of finite rank. Let Proj ($\mathfrak{T}$) denote the sub-category of $\mathfrak{V}$ whose objects are images of projection operators in $\mathfrak{T}$, and let Proj ($\mathfrak{F}$) $\subset \mathfrak{M}$ be defined similarly. Then it follows at once that Γ induces an equivalence of categories

$$\mathrm{Proj}\,(\mathfrak{T}) \longrightarrow \mathrm{Proj}\,(\mathfrak{F}) \quad .$$

But, by (1.4.14), Proj ($\mathfrak{T}$) = $\mathfrak{V}$. By definition Proj ($\mathfrak{F}$) is the category of finitely-generated projective $C(X)$-modules. Thus we have established the following:

PROPOSITION. Γ <u>induces an equivalence between the category of vector bundles over</u> X <u>and the category of finitely-generated projective modules over</u> $C(X)$.

§1.5. Additional structures. In linear algebra one frequently considers vector spaces with some additional structure, and we can do the same for vector bundles. For example we have already discussed hermitian metrics. The next most obvious example is to consider non-degenerate bilinear forms. Thus if V is a vector bundle a non-degenerate bilinear form on V means an element T of $HOM(V \otimes V, 1)$ which induces a non-degenerate element of $Hom(V_x \otimes V_x, C)$ for all $x \in X$. Equivalently T may be regarded as an element of $ISO(V, V^*)$. The vector bundle V together with this isomorphism T will be called a self-dual bundle.

If T is symmetric, i.e., if T_x is symmetric for all $x \in X$, we shall call (V, T) an orthogonal bundle. If T is skew-symmetric, i.e., if T_x is skew-symmetric for all $x \in X$, we shall call (V, T) a symplectic bundle.

Alternatively we may consider pairs (V, T) with $T \in ISO(V, \overline{V})$, where $\overline{V}$ denotes the complex conjugate bundle of V (obtained by applying the "complex conjugate functor" to V). Such a (V, T) may be called a self-conjugate bundle. The isomorphism T may also be thought of as an anti-linear isomorphism $V \to V$. As such we may form T^2. If T^2 = identity we may call (V, T) a real bundle. In fact the subspace $W \subset V$ consisting of all $v \in V$ with $Tv = v$ has the structure of a real vector bundle and V may be identified with $W \otimes_R C$, the

<u>complexification</u> of W. If T^2 = - identity then we may call (V, T) a <u>quaternion</u> bundle. In fact, we can define a quaternion vector space structure on each V_x by putting $j(v) = Tv$ the quaternions are generated over R by i, j with $ij = -ji$, $i^2 = j^2 = -1$).

Now if V has a hermitian metric h then this gives an isomorphism $\overline{V} \to V^*$ and hence turns a self-conjugate bundle into a self-dual one. We leave it as an exercise to the reader to examine in detail the symmetric† and skew-symmetric cases and to show that, up to homotopy, the notions of self-conjugate, orthogonal, symplectic, are essentially equivalent to self-dual, real, quaternion. Thus we may pick which ever alternative is more convenient at any particular stage. For example, the result of the preceeding sections extend immediately to real and quaternion vector bundles although the extension of (1.4.3) for example to orthogonal or sympletic bundles is not so immediate. On the other hand the properties of tensor products are more conveniently dealt with in the framework of bilinear forms. Thus the tensor product of (V, T) and (W, S) is $(V \otimes W, T \otimes S)$ and the symmetry properties of $T \otimes S$ follow at once from those of T and S. Note in particular that the tensor product of two symplectic bundles is orthogonal.

---

† The point is that GL(n, R) and O(n, C) have the same maximal compact subgroup O(n, R). Similar remarks apply in the skew case.

A self-conjugate bundle is a special case of a much more general notion. Let F, G be two continuous functors on vector spaces. Then by an $F \to G$ bundle we will mean a pair (V, T) where V is a vector bundle and $T \in ISO(F(V), G(V))$. Obviously a self-conjugate bundle arises by taking F = identity, $G = {}^*$. Another example of some importance is to take F and G to be multiplication by a fixed integer m, i.e.,

$$F(V) = G(V) = V \oplus V \oplus \cdots \oplus V \qquad \text{(m times)}.$$

Thus an $m \to m$ bundle (or more briefly an m-bundle) is a pair (V, T) where $T \in Aut(mV)$. The m-bundle (V, T) is <u>trivial</u> if there exists $S \in Aut(V)$ so that $T = mS$.

In general for $F \to G$ bundles the analogue of (1.4.3) does not hold, i.e., homotopy does not imply isomorphism. Thus the good notion of equivalence must incorporate homotopy. For example, two m-bundles (V_0, T_0) and (V_1, T_1) will be called <u>equivalent</u> if there is an m-bundle (W, S) on $X \times I$ so that

$$(V_i, T_i) \cong (W, S)|X \times \{i\}, \qquad i = 0, 1.$$

<u>Remark:</u> An m-bundle over K should be thought of as a "mod m vector bundle" over S(X).

§ 1.6. G-bundles over G-spaces. Suppose that G is a topological group. Then by a G-space we mean a topological space X together with a given continuous action of G on X, i.e., G acts on X and the map $G \times X \to X$ is continuous. A G-map between G-spaces is a map commuting with the action of G. A G-space E is a G-vector bundle over the G-space X if

(i) E is a vector bundle over X,

(ii) the projection $E \to X$ is a G-map,

(iii) for each $g \in G$ the map $E_x \to E_{g(x)}$ is a vector space homomorphism.

If G is the group of one element then of course every space is a G-space and every vector bundle is a G-vector bundle. At the other extreme if X is a point then X is a G-space for all G and a G-vector bundle over X is just a (finite-dimensional) representation space of G. Thus G-vector bundles form a natural generalization including both ordinary vector bundles and G-modules. Much of the theory of vector bundles over compact spaces generalizes to G-vector bundles provided G is also compact. This however, presupposes the basic facts about representations of compact groups. For the present, therefore we restrict ourselves to finite groups where no questions of analysis are involved.

There are two extreme kinds of G-space:

(i) X is a free G-space if $g \neq 1 \implies g(x) \neq x$,

(ii) X is a trivial G-space if $g(x) = x$ for all $x \in X$, $g \in G$,

We shall examine the structure of G-vector bundles in these two extreme cases.

Suppose then that X is a free G-space and let X/G be the orbit space. Then $\pi : X \to X/G$ is a finite covering map. Let E be a G-vector bundle over X. Then E is necessarily a free G-space. The orbit space E/G has a natural vector bundle structure over X/G : in fact $E/G \to X/G$ is locally isomorphic to $E \to X$ and hence the local triviality of E implies that of E/G. Conversely, suppose V is a vector bundle over X/G. Then π^*V is a G-vector bundle over X ; in fact, $\pi^*V \subset X \times V$ and G acts on $X \times V$ by $g(x, v) = (g(x), v)$. It is clear that $E \to E/G$ and $V \to \pi^*V$ are inverse functors. Thus we have

PROPOSITION 1.6.1. *If X is G-free G-vector bundles over X correspond bijectively to vector bundles over X/G by $E \to E/G$.*

Before discussing trivial G-spaces let us recall the basic fact about representations of finite groups, namely that there exists a finite set $V_1, \cdots, V_k$ of irreducible representations of G so that any representation V of G is isomorphic to a unique direct sum $\Sigma_{i=1}^k n_i V_i$. Now for any two G-modules (i.e., representation spaces) V, W we can define the vector space $\mathrm{Hom}_G(V, W)$ of G-homomorphisms. Then we have

$$\begin{aligned} \mathrm{Hom}_G(V_i, V_j) &= 0 & i \neq j \\ &\cong C & i = j \ . \end{aligned}$$

Hence for any V it follows that the natural map

$$\sum V_i \otimes \mathrm{Hom}_G(V_i, V) \longrightarrow V$$

is a G- isomorphism. In this form we can extend the result to G-bundles over a trivial G-space. In fact, if E is any G-bundle over the trivial G-space X we can define the homomorphism $Av \in \mathrm{END}\, E$ by

$$Av(e) = \frac{1}{|G|} \sum_{g \in G} g(e) \qquad e \in E$$

where $|G|$ denotes the order of G (This depends on the fact that, X being G-trivial, each $g \in G$ defines an endomorphism of E). It is immediate that Av is a projection operator for E and so its image, the invariant subspace of E, is a vector bundle. We denote this by E^G and call it the invariant sub-bundle of E. Thus if E, F are two G-bundles then $\mathrm{Hom}_G(E, F) = (\mathrm{Hom}(E, F))^G$ is again a vector bundle. In particular taking E to be the trivial bundle $\mathbf{V}_i = X \times V_i$ with its natural G-action we can consider the natural bundle map

$$\sum_{i=1}^{k} \mathbf{V}_i \otimes \mathrm{Hom}_G(\mathbf{V}_i, F) \longrightarrow F \ .$$

We have already observed that for a G-module F this is a G-isomorphism. In other words for any G-bundle F over X this is a G-isomorphism for all $x \in X$. Hence it is an isomorphism of G-bundles. Thus every G-bundle F is isomorphic to a G-bundle of the form $\Sigma \mathbf{V}_i \otimes E_i$ where E_i is a vector bundle with trivial G-action. Moreover the E_i are unique up to isomorphism. In fact we have

$$\begin{aligned} \mathrm{Hom}_G(\mathbf{V}_i, F) &\cong \sum_{j=1}^{k} \mathrm{Hom}_G(\mathbf{V}_i, \mathbf{V}_j \otimes E_j) \\ &\cong \sum_{j=1}^{k} \mathrm{Hom}_G(\mathbf{V}_i, \mathbf{V}_j) \otimes E_j \\ &\cong E_i . \end{aligned}$$

Thus we have established

PROPOSITION 1.6.2. Let X be a trivial G-space, $V_1, \cdots, V_k$ a complete set of irreducible G-modules, $\mathbf{V}_i = X \times V_i$ the corresponding G-bundles. Thus every G-bundle F over X is isomorphic to a direct sum $\Sigma_{i=1}^{k} \mathbf{V}_i \otimes E_i$ where the E_i are vector bundles with trivial G-action. Moreover the E_i are unique up to isomorphism and are given by $E_i = \mathrm{Hom}_G(\mathbf{V}_i, F)$.

We return now to the case of a general (compact) G-space X and we shall show how to extend the results of §1.4 to G-bundles.

Observe first that, if E is a G-bundle, G acts naturally on $\Gamma(E)$ by

$$(gs)(x) = g(s(g^{-1}(x))) \qquad s \in \Gamma(E) .$$

A section s is invariant if gs = g for all $g \in G$. The set of all invariant sections forms a subspace $\Gamma(E)^G$ of $\Gamma(E)$. The averaging operator

$$Av = \frac{1}{|G|} \sum g$$

defines as usual a homomorphism $\Gamma(E) \to \Gamma(E)^G$ which is the identity on $\Gamma(E)^G$.

LEMMA 1.6.3. Let X be a compact G-space $Y \subset X$ a closed sub G-space (i.e., invariant by G) and let E be a G-bundle over X. Then any invariant section $s : Y \to E|Y$ extends to an invariant section over X.

Proof: By (1.4.1) we can extend s to some section t of E over X. Then Av(t) is an invariant section of E over X, while over Y we have

$$Av(t) = Av(s) = s$$

since s is invariant. Thus Av(t) is the required extension.

If E, F are two G-bundles then $Hom(E, F)$ is also a G-bundle and we have

$$\Gamma(Hom(E,F))^G \cong HOM_G(E, F) .$$

Hence the G-analogues of (1.4.2) and (1.4.3) follow at once from (1.6.3). Thus we have

LEMMA 1.6.4. _Let_ Y _be a compact_ G-_space_, X _a_ G-_space_, $f_t : Y \to X$ $(0 \leq t \leq 1)$ _a_ G-_homotopy and_ E _a_ G-_vector bundle over_ X. _Then_ f_0^*E _and_ f_1^*E _are isomorphic_ G-_bundles._

A G-homotopy means of course a G-map $F : Y \times I \to X$ where I is the unit interval with trivial G-action. A G-space is G-contractible if it is G-homotopy equivalent to a point. In particular, the cone over a G-space is always G-contractible. By a _trivial_ G-bundle we shall mean a G-bundle isomorphic to a product $X \times V$ where V is a G-module. With these definitions (1.4.4) – (1.4.11) extend without change to G-bundles. We have only to observe that if h is a metric for E then $Av(h)$ is an invariant metric.

To extend (1.4.12) we observe that if $V \subset \Gamma(E)$ is ample then $\Sigma_{g \in G}\, gV \subset \Gamma(E)$ is ample and invariant. This leads at once to the appropriate extension of (1.4.14).

In extending (1.4.15) we have to consider Grassmannians of G-subspaces of $m\, \Sigma_{i=1}^{k} V_i$ for $m \to \infty$, where as before

$V_1, \cdots, V_k$ denote a complete set of irreducible G-modules. We leave the formulation to the reader.

Finally, consider the module interpretation of vector bundles. Write $A = C(X)$. Then if X is a G-space G acts on A as a group of algebra automorphisms. If E is a G-vector bundle over X then $\Gamma(E)$ is a projective A-module and G acts on $\Gamma(E)$, the relation between the A - and G - actions being

$$g(as) = g(a)g(s) \qquad a \in A,\ g \in G,\ s \in \Gamma(E).$$

We can look at this another way if we introduce the "twisted group algebra" B of G over A, namely elements of B are linear combinations $\Sigma_{g\in G}\, a_g g$ with $a_g \in A$ and the product is defined by

$$(ag)(a'g') = (ag(a'))gg' .$$

In fact, $\Gamma(E)$ is then just a B-module. We leave it as an exercise to the reader to show that the category of G-vector bundles over X is equivalent to the category of B-modules which are finitely generated and projective over A.

CHAPTER II. K-Theory

§ 2.1. Definitions. If X is any space, the set $\mathrm{Vect}(X)$ has the structure of an abelian semigroup, where the additive structure is defined by direct sum. If A is any abelian semigroup, we can associate to A an abelian group $K(A)$ with the following property: there is a semigroup homomorphism $\alpha : A \to K(A)$ such that if G is any group, $\gamma : A \to G$ any semigroup homomorphism, there is a unique homomorphism $\chi : K(A) \to G$ such that $\gamma = \chi\alpha$. If such a $K(A)$ exists, it must be unique.

The group $K(A)$ is defined in the usual fashion. Let $F(A)$ be the free abelian group generated by the elements of A, let $E(A)$ be the subgroup of $F(A)$ generated by those elements of the form $a + a' - (a \oplus a')$, where $\oplus$ is the addition in A, $a, a' \in A$. Then $K(A) = F(A)/E(A)$ has the universal property described above, with $\alpha : A \to K(A)$ being the obvious map.

A slightly different construction of $K(A)$ which is sometimes convenient is the following. Let $\Delta : A \to A \times A$ be the diagonal homomorphism of semi-groups, and let $K(A)$ denote the set of cosets of $\Delta(A)$ in $A \times A$. It is a quotient semi-group, but the interchange of factors in $A \times A$ induces an inverse in $K(A)$ so that $K(A)$ is a group. We then define $\alpha_A : A \to K(A)$ to be the composition of $a \to (a, 0)$ with the natural projection $A \times A \to K(A)$ (we assume

A has a zero for simplicity). The pair $(K(A), \alpha_A)$ is a functor of A so that if $\gamma : A \to B$ is a semi-group homomorphism we have a commutative diagram

$$\begin{array}{ccc} A & \xrightarrow{\alpha_A} & K(A) \\ \gamma \downarrow & & \downarrow K(\gamma) \\ B & \xrightarrow{\alpha_B} & K(B) \end{array}$$

If B is a group α_B is an isomorphism. That shows $K(A)$ has the required universal property.

If A is also a semi-ring (that is, A possesses a multiplication which is distributative over the addition of A) then $K(A)$ is clearly a ring.

If X is a space, we write $K(X)$ for the ring $K(\mathrm{Vect}(X))$. No confusion should result from this notation. If $E \in \mathrm{Vect}(X)$, we shall write $[E]$ for the image of E in $K(X)$. Eventually, to avoid excessive notation, we may simply write E instead of $[E]$ when there is no danger of confusion.

Using our second construction of K it follows that, if X is a space, every element of $K(X)$ is of the form $[E] - [F]$, where

E, F are bundles over X. Let G be a bundle such that $F \oplus G$ is trivial. We write $\underline{n}$ for the trivial bundle of dimension n. Let $F \oplus G = \underline{n}$. Then $[E] - [F] = [E] + [G] - ([F] + [G]) = [E \oplus G] - [\underline{n}]$ Thus, every element of K(X) is of the form $[H] - [\underline{n}]$.

Suppose that E, F are such that $[E] = [F]$, then again from our second construction of K it follows that there is a bundle G such that $E \oplus G \cong F \oplus G$. Let G' be a bundle such that $G \oplus G' \cong \underline{n}$. Then $E \oplus G \oplus G' \cong F \oplus G \oplus G'$, so $E \oplus \underline{n} \cong F \oplus \underline{n}$. If two bundles become equivalent when a suitable trivial bundle is added to each of them, the bundles are said to be stably equivalent. Thus, $[E] = [F]$ if and only if E and F are stably equivalent.

If we define $\mathrm{Vect}_n(X) \to \mathrm{Vect}_{n+1}(X)$ by the addition of a trivial line-bundle it follows that we have

LEMMA 2.1.1. *For a compact connected space* X,

$$K(X) \cong Z \times \varinjlim_{n} \mathrm{Vect}_n(X)$$

Suppose $f : X \to Y$ is a continuous map. Then $f^* : \mathrm{Vect}(Y) \to \mathrm{Vect}(X)$ induces a ring homomorphism $f^* : K(Y) \to K(X)$. By (1.4.3) this homomorphism depends only on the homotopy class of f.

We conclude this section by giving the homotopy-theoretic definition of K(X). This is essentially a re-interpretation of the results on $\mathrm{Vect}_n(X)$ in Chapter I. We introduce the "infinite" unitary group U defined as

$$U = \varinjlim_{n} U(n)$$

under the standard inclusions. Correspondingly for the classifying spaces we have

$$BU = \varinjlim_n BU(n) .$$

Recall also that the classifying space $BU(n)$ can be taken to be the limit Grassmannian:

$$BU(n) = \varinjlim_m G_n(C^m) .$$

Hence from Theorem 1.4.15, for any compact connected X, we have

$$\begin{aligned}[X,BU] &\cong \varinjlim_n [X,BU(n)] \\ &\cong \varinjlim_n \varinjlim_m [X,G_n(C^m)] \\ &\cong \varinjlim_n \mathrm{Vect}_n(X) \\ &\cong \tilde{K}(X) \qquad \text{by} \quad (2.1.1)\end{aligned}$$

Equivalently, for all compact X, we have

PROPOSITION 2.1.10: $K(X) \cong [X, Z \times BU]$.

Similarly, (2.17) can be written as

PROPOSITION 2.1.11: $\tilde{K}(SX) \cong [X,U]$.

In fact 2.1.11 is a consequence of 2.1.10 and the homotopy equivalence:

$$U \sim \Omega BU .$$

More generally we have for all $n \geq 1$.

PROPOSITION 2.1.12: $K^{-n}(X) \cong [X,\Omega^n BU] \cong [X,\Omega^{n-1}U]$

§2.2. Elementary Properties. We next define relative groups $K(X,Y)$ for a compact pair (X,Y), with $Y \subset X$.

Let C denote the category of compact spaces, C^+ the category of compact spaces with distinguished basepoint, and C^2 the category of compact pairs. We define functors:

$$C^2 \longrightarrow C^+$$
$$C \longrightarrow C^2$$

by sending a pair (X, Y) to X/Y with basepoint Y/Y (if $Y \neq \emptyset$, the empty set, X/Y is understood to be the disjoint union of X with a point.) We send a space X to the pair $(X, \emptyset)$. The composite $C \to C^+$ is given by $X \to X^+$, where X^+ denotes $X/\emptyset$.

If X is in C^+, we define $\tilde{K}(X)$ to be the kernel of the map $i^* : K(X) \to K(x_0)$ where $i : x_0 \to X$ is the inclusion of the basepoint. If $c : X \to x_0$ is the collapsing map then c^* induces a splitting $K(X) \cong \tilde{K}(X) \oplus K(x_0)$. This splitting is clearly natural for maps in C^+. Thus $\tilde{K}$ is a functor on C^+. Also, it is clear that $K(X) \cong \tilde{K}(X^+)$. We define $K(X, Y)$ by $K(X,Y) = \tilde{K}(X/Y)$. In particular $K(X, \emptyset) \cong K(X)$. Since $\tilde{K}$ is a functor on C^+ it follows that $K(X, Y)$ is a contravariant functor of (X, Y) in C^2.

We now introduce the "smash product" operation in C^+. If $X, Y \in C^+$ we put

$$X \wedge Y = X \times Y / X \vee Y$$

where $X \vee Y = X \times y_0 \cup x_0 \times Y$, x_0, y_0 being the base-points of X, Y respectively. For any three spaces $X, Y, Z \in C^+$ we have a natural homeomorphism

$$X \wedge (Y \wedge Z) \approx (X \wedge Y) \wedge Z$$

and we shall identify these spaces by the homeomorphism.

Let I denote the unit interval $[0, 1]$ and let $\partial I = \{0\} \cup \{1\}$ be its boundary. We take $I/\partial I \in C^+$ as our standard model of the circle S^1. Similarly if I^n denotes the unit cube in R^n we take $I^n/\partial I^n$ as our model of the n-sphere S^n. Then we have a natural homeomorphism

$$S^n \approx S^1 \wedge S^1 \wedge \cdots \wedge S^1 \qquad \text{(n factors)} .$$

For $X \in C^+$ the space $S^1 \wedge X \in C^+$ is called the <u>reduced suspension</u> of X, and often written as SX. The n-th iterated suspension $SS \cdots SX$ (n times) is naturally homeomorphic to $S^n \wedge X$ and is written briefly as S^nX.

DEFINITION 2.2.2. For $n \geq 0$

$$\tilde{K}^{-n}(X) = \tilde{K}(S^n X) \qquad \text{for } X \in \mathcal{C}^+$$

$$K^{-n}(X, Y) = \tilde{K}^{-n}(X/Y) = \tilde{K}(S^n(X/Y)) \qquad \text{for } (X, Y) \in \mathcal{C}^2$$

$$K^{-n}(X) = K^{-n}(X, \emptyset) = \tilde{K}(S^n(X^+)) \qquad \text{for } X \in \mathcal{C} .$$

It is clear that all these are contravariant functors on the appropriate categories.

Before proceeding further we define the cone on X by

$$CX = I \times X/\{0\} \times X .$$

Thus C is a functor $C : \mathcal{C} \to \mathcal{C}^+$. We identify X with the subspace $\{1\} \times X$ of CX. The space $CX/X = I \times X/\partial I \times X$ is called the unreduced suspension of X. Note that this is a functor $\mathcal{C} \to \mathcal{C}^+$ whereas the reduced suspension S is a functor $\mathcal{C}^+ \to \mathcal{C}^+$. If $X \in \mathcal{C}^+$ with base-point x_0 then we have a natural inclusion map

$$I \approx Cx_0/x_0 \longrightarrow CX/X$$

and the quotient space obtained by collapsing I in CX/X is just SX. Thus by (1.4.8) $p : CX/X \to SX$ induces an isomorphism $K(SX) \cong K(CX/X)$ and hence also an isomorphism $\tilde{K}(SX) \cong K(CX, X)$. Thus the use of SX for both the reduced and unreduced suspensions leads to no problems .

If $(X, Y) \in \mathcal{C}^2$ we define $X \cup CY$ to be the space obtained from X and CY by identifying the subspaces $Y \subset X$ and $\{1\} \times Y \subset CY$. Taking the base-point of CY as base-point of $X \cup CY$ we have

$$X \cup CY \in \mathcal{C}^+ .$$

We note that X is a subspace of $X \cup CY$ and that there is a natural homeomorphism

$$X \cup CY/X \approx CY/Y .$$

Thus, if $Y \in \mathcal{C}^+$,

$$\begin{aligned} K(X \cup CY, X) &\cong K(CY, Y) \\ &\cong \tilde{K}(SY) \\ &= \tilde{K}^{-1}(Y) . \end{aligned}$$

Now we begin with a simple lemma

LEMMA 2.2.3. <u>For</u> $(X, Y) \in \mathcal{C}^2$ <u>we have an exact sequence</u>

$$K(X, Y) \xrightarrow{j^*} K(X) \xrightarrow{i^*} K(Y)$$

<u>where</u> $i : Y \to X$ <u>and</u> $j : (X, \emptyset) \to (X, Y)$ <u>are the inclusions.</u>

<u>Proof</u>: The composition i^*j^* is induced by the composition $ji : (Y, \emptyset) \to (X, Y)$, and so factors through the zero group $K(Y, Y)$. Thus $i^*j^* = 0$. Suppose now that $\xi \in \operatorname{Ker} i^*$. We may represent ξ in the form $[E] - [n]$ where E is a vector bundle over X. Since $i^*\xi = 0$ it follows that $[E|Y] = [n]$ in $K(Y)$. This implies that for some integer m we have

$$(E \oplus m)|Y = n \oplus m$$

i.e., we have a trivialization α of $(E \oplus m)|Y$. This defines a bundle $E \oplus m/\alpha$ on X/Y and so an element

$$\eta = [E \oplus m/\alpha] - [n \oplus m] \in \widetilde{K}(X/Y) = K(X, Y) \quad .$$

Then

$$\begin{aligned} j^*(\eta) &= [E \oplus m] - [n \oplus m] \\ &= [E] - [n] = \xi \quad . \end{aligned}$$

Thus $\operatorname{Ker} i^* = \operatorname{Im} j^*$ and the exactness is established.

COROLLARY 2.2.4. <u>If</u> $(X, Y) \in \mathcal{C}^2$ <u>and</u> $Y \in \mathcal{C}^+$ <u>(so that, taking the same base-point of</u> X, <u>we have</u> $X \in \mathcal{C}^+$ <u>also)</u>, <u>then the sequence</u>

$$K(X, Y) \longrightarrow \widetilde{K}(X) \longrightarrow \widetilde{K}(Y)$$

<u>is exact</u>.

<u>Proof:</u> This is immediate from (2.2.3) and the natural isomorphisms

$$K(X) \cong \tilde{K}(X) \oplus K(y_0)$$

$$K(Y) \cong \tilde{K}(Y) \oplus K(y_0) \quad .$$

We are now ready for our main proposition:

PROPOSITION 2.2.5. <u>For</u> $(X, Y) \in C^+$ <u>there is a natural exact sequence (infinite to the left)</u>

$$\cdots K^{-2}(Y) \xrightarrow{\delta} K^{-1}(X, Y) \xrightarrow{j^*} K^{-1}(X) \xrightarrow{i^*} K^{-1}(Y) \xrightarrow{\delta} K^0(X, Y)$$
$$\xrightarrow{j^*} K^0(X) \xrightarrow{i^*} K^0(Y) \quad .$$

<u>Proof:</u> First we observe that it is sufficient to show that, for $(X, Y) \in C^2$ and $Y \in C^+$, we have an exact sequence of five terms

$$(*) \quad \tilde{K}^{-1}(X) \xrightarrow{i^*} \tilde{K}^{-1}(Y) \xrightarrow{\delta} \tilde{K}^0(X, Y) \xrightarrow{j^*} \tilde{K}^0(X) \xrightarrow{i^*} \tilde{K}^0(Y) \quad .$$

In fact, if this has been established then, replacing (X, Y) by $(S^n X, S^n Y)$ for $n = 1, 2, \cdots$ we obtain an infinite sequence continuing $(*)$. Then replacing (X, Y) by (X^+, Y^+) where (X, Y) is any pair in C^2 we get the infinite sequence of the enunciation. Now (2.2.4) gives the exactness of the last three terms of $(*)$. To get exactness at the remaining places we shall apply (2.2.4) in turn to the pairs $(X \cup CY, X)$ and $((X \cup CY) \cup CX, X \cup CY)$. First, taking the pair $(X \cup CY, X)$ we get an exact sequence (where k, m are the natural inclusions)

$$K(X \cup CY, X) \xrightarrow{m^*} \tilde{K}(X \cup CY) \xrightarrow{k^*} \tilde{K}(X) \ .$$

Since CY is contractible 1.4.8) implies that

$$p^* : \tilde{K}(X/Y) \longrightarrow \tilde{K}(X \cup CY)$$

is an isomorphism where

$$p : X \cup CY \longrightarrow X \cup CY/CY = X/Y$$

is the collapsing map. Also the composition $k^* p^*$ coincides with j^*. Let

$$\theta : K(X \cup CY, X) \longrightarrow K^{-1}(Y)$$

be the isomorphism introduced earlier. Then defining

$$\delta : K^{-1}(Y) \longrightarrow K(X, Y)$$

by $\delta = m^* \theta^{-1}$ we obtain the exact sequence

$$\tilde{K}^{-1}(Y) \xrightarrow{\delta} K(X, Y) \xrightarrow{j^*} \tilde{K}(X)$$

which is the middle part of (*).

Finally, we apply (2.2.4) to the pair

$$(X \cup C_1Y \cup C_2X,\ X \cup C_1Y)$$

where we have labelled the cones C_1 and C_2 in order to distinguish between them. (see figure).

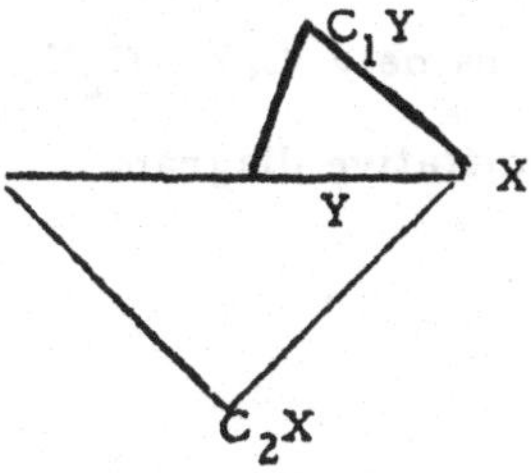

Thus we obtain an exact sequence

$$K(X \cup C_1Y \cup C_2X,\ X \cup C_1Y) \longrightarrow \tilde{K}(X \cup C_1Y \cup C_2X) \longrightarrow \tilde{K}(X \cup C_1Y) \quad .$$

It will be sufficient to show that this sequence is isomorphic to the sequence obtained from the first three terms of (*). In view of the definition of δ it will be sufficient to show that the diagram

$$\text{(D)} \qquad \begin{array}{ccc} K(X \cup C_1Y \cup C_2X,\ X \cup C_1Y) & \longrightarrow & \tilde{K}(X \cup C_1Y \cup C_2X) \\ \| & & \| \\ \tilde{K}(C_2X/X) & & \tilde{K}(C_1Y/Y) \\ \| & & \| \\ K^{-1}(X) & \xrightarrow{i^*} & K^{-1}(Y) \end{array}$$

commutes up to sign. The difficulty lies, of course, in the fact that i^* is induced by the inclusion

$$C_2Y \longrightarrow C_2X$$

and that in the above diagram we have C_1Y and not C_2Y. To deal with this situation we introduce the double cone on Y namely $C_1Y \cup C_2Y$. This fits into the commutative diagram of maps

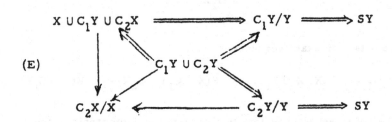

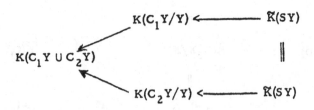

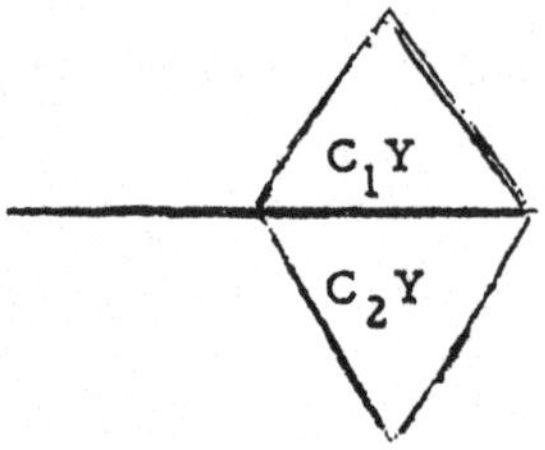

$$\text{(E)}\qquad \begin{array}{ccccc} X \cup C_1Y \cup C_2X & \Longrightarrow & C_1Y/Y & \Longrightarrow & SY \\ \downarrow & \nwarrow \quad & \nearrow & & \\ & & C_1Y \cup C_2Y & & \\ & \swarrow \quad & \searrow & & \\ C_2X/X & \longleftarrow & C_2Y/Y & \Longrightarrow & SY \end{array}$$

where all double arrows $\Longrightarrow$ induce isomorphism in K. Using this diagram we see that diagram (D) will commute up to sign provided the diagram induced by (E)

$$\begin{array}{ccccc} & & K(C_1Y/Y) & \longleftarrow & \tilde{K}(SY) \\ & \swarrow & & & \| \\ K(C_1Y \cup C_2Y) & & & & \\ & \nwarrow & & & \\ & & K(C_2Y/Y) & \longleftarrow & \tilde{K}(SY) \end{array}$$

commutes up to sign. This will follow at once from the following lemma which is in any case of independent interest and will be needed later

LEMMA 2.2.6. Let $T : S^1 \to S^1$ be defined by $T(t) = 1 - t$, $t \in I$ (we recall that $S^1 = I/\partial I$) and let $T \wedge 1 : SY \to SY$ be the map induced by T on S^1 and the identity on Y (for $Y \in C^+$). Then $(T \wedge 1)^* y = -y$ for $y \in \tilde{K}(SX)$.

This lemma in turn is an easy corollary of the following:

LEMMA 2.2.7. For any map $f : Y \to GL(n, C)$ let E_f denote the corresponding vector bundle over SY. Then $f \to [E_f] - [n]$ induces a group isomorphism

$$\lim_{n \to \infty} [Y, GL(n, C)] \cong \tilde{K}(SY)$$

where the group structure on the left is induced from that of $GL(n,C)$.

In fact, the operation $(T \wedge 1)^*$ on $\tilde{K}(SY)$ corresponds by the isomorphism of (2.2.7) to the operation of replacing the map $y \to f(y)$ by $y \to f(y)^{-1}$, i.e., it corresponds to the inverse in the group. Thus (2.2.7) implies (2.2.6) and hence (2.2.5). It remains therefore to establish (2.2.7). Now (1.4.9) implies that $f \to [E_f] - [n]$ induces a bijection of sets

$$\lim [Y, GL(n, C)] \longrightarrow \tilde{K}(SY) \quad .$$

The fact that this is in fact a group homomorphism follows from the homotopy connecting the two maps $GL(n) \times GL(n) \to GL(2n)$ given by

$$A \times B \longrightarrow \begin{pmatrix} A & 0 \\ 0 & B \end{pmatrix}$$

and

$$A \times B \longrightarrow \begin{pmatrix} AB & 0 \\ 0 & 1 \end{pmatrix} \quad .$$

This homotopy is given explicitly by

$$\rho_t(A \times B) = \begin{pmatrix} A & 0 \\ 0 & 1 \end{pmatrix} \begin{pmatrix} \cos t & \sin t \\ -\sin t & \cos t \end{pmatrix} \begin{pmatrix} 1 & 0 \\ 0 & B \end{pmatrix} \begin{pmatrix} \cos t & \sin t \\ \sin t & \cos t \end{pmatrix}$$

where $0 \leq t \leq \pi/2$.

From (2.2.5) we deduce at once:

COROLLARY 2.2.8. If Y is a retract of X, then for all $n \geq 0$, the sequence $K^{-n}(X,Y) \to K^{-n}(X) \to K^{-n}(Y)$ is a split short exact sequence, and

$$K^{-n}(X) \cong K^{-n}(X,Y) \oplus K^{-n}(Y) \quad .$$

COROLLARY 2.2.9. If X, Y are two spaces with basepoints, the projection maps $\pi_X : X \times Y \to X$, $\pi_Y : X \times Y \to Y$ induce an isomorphism for all $n \geq 0$

$$\widetilde{K}^{-n}(X \times Y) \cong \widetilde{K}^{-n}(X \wedge Y) \oplus \widetilde{K}^{-n}(X) \oplus \widetilde{K}^{-n}(Y) \quad .$$

<u>Proof:</u> X is a retract of $X \times Y$, and Y is a retract of $(X \times Y)/X$. The result follows by two applications of (2.2.8).

Since $\tilde{K}^0(X \wedge Y)$ is the kernel of $i_X^* \oplus i_Y^* : K^0(X \times Y) \to K^0(X) \oplus K^0(Y)$, the usual tensor product $K^0(X) \otimes K^0(Y) \to K^0(X \times Y)$ induces a pairing $\tilde{K}^0(X) \otimes \tilde{K}^0(Y) \to \tilde{K}^0(X \wedge Y)$. Thus, we have a pairing

$$\tilde{K}^{-n}(X) \otimes \tilde{K}^{-m}(Y) \longrightarrow \tilde{K}^{-n-m}(X \wedge Y),$$

since $S^n X \wedge S^m Y = S^n \wedge S^m \wedge X \wedge Y = S^{n+m} \wedge X \wedge Y$. Replacing X by X^+, Y by Y^+, we have

$$K^{-n}(X) \otimes K^{-m}(Y) \longrightarrow K^{-n-m}(X \times Y) .$$

§2.3. <u>The Bott Periodicity Theorem</u>

The fundamental theorem of K-theory is the Bott periodicity theorem which asserts that we have a natural isomorphism

(a) $K(X) \cong K^{-2}(X)$.

Based on this periodicity property we can then extend the definition of $K^{-n}(X)$ to all integers n, so as to be periodic mod 2, and we will get a full cohomology-type theory. With this machinery we will then have a powerful new tool for attacking many problems in algebraic topology.

Before we embark on the proof of the periodicity theorem it may be helpful if we digress to give a few historical remarks and make some comments on the different proofs that have been given.

The original approach of Bott [1] consisted of studying the loop space $\Omega U(n)$ by means of Morse Theory. For the limit group U, Bott established a homotopy equivalence

(b) $U \sim \Omega^2 U$.

In view of Proposition 2.2.7 we see that (b) implies (a) for all X. The converse is also true as we see by taking X to be a compact approximation to U.

In view of the homotopy equivalence $U \sim \Omega BU$, the equivalence (b) can be replaced by

(c) $\Omega U \sim Z \times BU$

and this is in fact what Bott proved. More precisely he constructed an explicit embedding

$$\beta_n : G_n(C^{2n}) = \frac{U(2n)}{U(n)\times U(n)} \to \Omega U(2n)$$

and showed that this gave the minimum for the "Energy" functional F on $\Omega U(2n)$. Moreover all other critical sets of F have a Morse index which goes to infinity with n. Applying the Morse theory and letting $n \to \infty$ Bott deduced (c) and hence (b).

The proof which we will give will be more direct in that we shall essentially construct two maps

$$\beta : Z \times BU \to \Omega U, \qquad \alpha : \Omega U \to Z \times BU$$

and show that they are homotopy inverses of each other. The map β is in effect the limit of Bott's β_n and is elementary. The main point is to construct α, and for this it is convenient to use another model of $Z \times BU$. As explained in the Appendix, if F is the space of Fredholm operators on a complex Hilbert Space H (of infinite-dimension) there is a map

$$\text{index} : [X,F] \to K(X)$$

which (using Kuiper's Theorem on the contractibility of the unitary group of H) is an isomorphism. This means that there is a homotopy equivalence

$$F \to Z \times BU ,$$

so that it is sufficient to construct α as a map

$$\alpha : \Omega U \to F .$$

Note that, by comparing ths map α, with the "index map" $F \to Z \times BU$, we get a map

$$\alpha' : \Omega U \to Z \times BU$$

which is what we are really after. Thus we do not need Kuiper's Theorem, only the existence of the index.

Now for any continuous map

$$f : S^1 \to U(n)$$

one can define a "Toeplitz operator" T_f, acting on the Hilbert space of holomorphic functions on the disc $|z| \leq 1$ with values in C^n. We recall that T_f consists in matrix multiplication by $f(z)$ followed by orthogonal projection onto the holomorphic functions (i.e. taking the positive part of the Fourier expansion). We then define

$$\alpha(f) = T_f$$

and show that this is consistent with increasing n.

A complete account of the periodicity theorem along these lines is given in [2] , and this is in many ways the optimal proof. However, it is possible to replace the Hilbert space theory by purely algebraic methods based on truncating the Fourier series. This leads to a proof which is in a sense more elementary and this is the proof which will now be presented in detail. To some extent it is the version given in [3] but the rather lengthy explicit verification of the homotopy $\beta\alpha \sim 1$ will be circumvented here by a simple trick introduced in [2]. In fact the explicit formulas in [3] were motivated by the requirements of elliptic boundary value problems, but they can be dispensed with for the purely topological theory.

After this lengthy pre-amble we return to give the formal treatment of the periodicity theorem in the form (a). We shall not work with the classifying spaces BU or F which were introduced here merely for the sake of comparison with Bott's formulation.

We shall identify the 2-sphere S^2 with the one-point compactification of C:

$$S^2 = C \cup \infty = P_1(C) ,$$

and ∞ will be regarded as the base-point. Now, on $P_n(C)$ there is the standard (Hopf) line-bundle H^* whose fibre at $y \in P_n(C)$ is the complex line in C^{n+1} represented by y. The dual bundle H is basic in algebraic geometry because H is holomorphic and its holomorphic sections are just the linear forms on C^{n+1}. In particular for $n = 1$ the bundle H can be constructed from trivial bundles E^0, E^∞ over the interior and exterior of the unit circle $|z| = 1$ with the identification

$$z : E^0 \to E^\infty$$

along $|z| = 1$. In view of Lemma 1.4.9 this means that H generates the multiplicative group $\mathrm{Vect}_1(S^2)$ of line-bundles over S^2. Put

$$b = [H] - [1] \in \tilde{K}(S^2) = K^{-2}(\text{point}) ,$$

and, for any X, let

$$\beta : K(X) \to K^{-2}(X)$$

denote multiplication by the element b. We shall refer to b as the Bott generator and β as the Bott homomorphism. The Bott periodicity theorem is now

THEOREM (2.3.1). $\beta : K(X) \to K^{-2}(X)$ is an isomorphism.

As indicated earlier we shall explicitly construct a homomorphism

$$\alpha : K^{-2}(X) \to K(X)$$

satisfying some basic properties, and from these we shall formally deduce that α is a 2-sided inverse of β. The construction of α involves some elementary algebra and we start therefore with a few algebraic preliminaries.

Let

$$p(z) = a_m z^m + a_{m-1} z^{m-1} + \dots + a_0$$

be a polynomial in z whose coefficients are complex $n\times n$-matrices. Then

$$d(z) = \det p(z) = (\det a_m) z^{mn} + \dots$$

is a polynomial of degree precisely mn provided $\det a_m \neq 0$, which we assume for the time being. We now regard $p(z)$ as defining a homomorphism, by matrix multiplication, of free

A-modules

$$A^n \overset{p}{\to} A^n$$

when $A = C[z]$ is the polynomial ring.

Since $d(z)p(z)^{-1}$ has coefficients in A it follows that

$$M(p) = \text{Coker } p$$

is an A-module which is annihilated by $d(z)$. Thus $M(p)$ is a torsion-module whose support is the set of roots of $d(z)$. Since a_m is invertible we have

$$z^m = -a_m^{-1}(a_{m-1}z^{m-1} + \ldots + a_0)$$

which shows that the elements

$$e_i z^j \qquad 1 \le i \le n \;, \qquad 0 \le j \le m-1$$

give a C-basis for M, where the e_i are the standard basis of C^n.

Multiplication by z gives a linear transformation on M which in this basis is represented by the block matrix

$$T = \begin{bmatrix} 0 & & & & & b_0 \\ 1 & 0 & & & & b_1 \\ & 1 & 0 & & & b_2 \\ & & \cdot & & & \cdot \\ & & & \cdot & & \cdot \\ & & & & \cdot & \cdot \\ & & & & 0 & b_{m-2} \\ & & & & 1 & b_{m-1} \end{bmatrix}$$

where $b_i = -a_m^{-1} a_i$. The A-module structure of M is entirely determined by T. Note that

$$\det(zI - T) = (\det a_m)^{-1} d(z)$$

so that the roots of $d(z)$, giving the support of M, are just the eigenvalues of T.

Suppose next that $d(z)$ has no roots on the unit circle $|z| = 1$, so that we can factorize it as

$$d(z) = d^+(z)d^-(z)$$

where the roots of $d^+(z)$ satisfy $|z| < 1$, while those of $d^-(z)$ satisfy $|z| > 1$. Clearly $d^\pm$ are unique up to a multiplicative constant and this can be fixed by requiring, for example that $d^+(1) = 1$.

The factorization of $d(z)$ defines a corresponding direct sum decomposition

$$(2.3.2) \qquad M(p) = M^+(p) \oplus M^-(p)$$

where $M^\pm(p)$ are respectively the kernels of multiplication by $d^\pm(z)$ on $M(p)$. To see this observe

(i) $M^+ \cap M = 0$ because d^+ and d^- have no common factor (so $1 = \alpha d^+ + \beta d^-$ with $\alpha, \beta \in A$) ,

(ii) $d^+(M^+) \subset M^-$ since $d^- d^+ M = dM = 0$, so that

$$\dim M^+ + \dim M^- \geq \dim M .$$

In terms of the matrix T we can also define M^+ as the image of the spectral projection

$$P^+ = \frac{1}{2\pi i} \int_{|z|=1} \frac{dz}{z-T} ,$$

as one easily sees by using the Jordan normal form of T.

If we now vary p continuously, still keeping our assumptions that $d(z)$ has no roots on $|z|=1$ and that $\det a_m \neq 0$, we see that d^+ will remain of constant degree and will vary continuously with p. Thus, as a homomorphism,

$$d^+ : M(p) \to M(p)$$

has constant rank and is continuous in p (i.e. it is a strict homomorphism in the terminology of Chapter I). Hence $M^+(p)$, its kernel, forms a vector bundle M^+ over the space of admissible p. This can also be seen from the continuity of the projection operator P^+.

This bundle will be the key element of our construction but before proceeding further we need to relax the assumption on the leading coefficient, and we now explain how to do this.

As long as $d(z) \not\equiv 0$ we can always form the torsion module $M(p)$, and if $d(z)$ has no roots on $|z| = 1$ we can still decompose it as in (2.3.2). However $\dim M(p)$ is no longer a locally constant function of p. Essentially if $\deg d(z) < mn$ then some roots have become "infinite" and we lose the corresponding eigenspaces. In particular the family of $M(p)$ does not form a vector bundle. However, the difficulty lies only in $M^-(p)$, while dim $M^+(p)$ will remain locally constant and the $M^+(p)$ will again form a vector bundle. Technically the way to verify this is to modify the definition of $M(p)$ to include the contribution from the "roots at infinity". This can be done as follows.

First pick α with $|\alpha| > 1$ and $d(\alpha) \neq 0$ and consider the birational transformation

$$(2.3.3) \qquad \omega = \frac{1-\bar{\alpha}z}{z-\alpha}\,, \qquad z = \frac{1+\alpha w}{w+\bar{\alpha}}$$

which preserves the unit circle and takes $z = \alpha$ to $w = \infty$. Define a new polynomial q by

$$(2.3.4) \qquad q(w) = (w+\bar{\alpha})^m\, p\left(\frac{1+\alpha w}{w+\bar{\alpha}}\right).$$

The leading coefficient of q is

$$a_m\alpha^m + a_{m-1}\alpha^{m-1} + \ldots = p(\alpha)$$

which is non-singular since $d(\alpha) \neq 0$. We can now introduce the module $M(q)$, with its standard basis, and proceed as before to decompose it as in (2.3.2). It is straightforward to check that we have a natural isomorphism

$$M^+(p) \cong M^+(q)$$

so that in this way we have given the family of $M^+(p)$ a

vector bundle structure. This works for all p with $d(\alpha) = \det p(\alpha) \neq 0$. By varying α we get a covering of the space of all admissible p (i.e. with $\det p(z) \neq 0$ on $|z| = 1$). So it remains only to check that the different choices of α give the same vector bundle structure, i.e. the same topology on the family of $M^+(p)$.

If we have values α, β with $d(\alpha) \neq 0$, $d(\beta) \neq 0$ we can by a preliminary change of variable, assume $\beta = \infty$. The transformation (2.3.3) will now induce an isomorphism $M(p) \cong M(q)$ where q is given by (2.3.4). The vector space $M(p)$ has its standard basis $T^j(e_i)$ (where T is multiplication by z) and $M(q)$ has its standard basis $S^j(e_i)$ (where S is multiplication by w). The isomorphism $M(p) \cong M(q)$ relative to these two standard tasks can be explicitly computed by using the functional relation

$$S = \frac{1-\bar{\alpha}T}{T-\alpha} .$$

The resulting matrix will clearly be rational in the coefficients of the original polynomial p and its denominator will be a power of $d(\alpha)$. This gives the required continuity.

We can sum up our results in the following proposition

<u>PROPOSITION (2.3.5).</u> <u>Let</u> $P(m,n)$ <u>denote the space of all polynomials</u>

$$p(z) = a_m z^m + \dots a_0$$

<u>with</u> $n \times n$-<u>matrices as coefficients and such that</u>

$$d(z) = \det p(z) \neq 0 \quad \text{on} \quad |z| = 1 .$$

<u>Let</u> $M(p)$ <u>be the cokernel of the homomorphism</u>

$$C[z]^n \xrightarrow{p} C[z]^n$$

<u>and let</u> $M^+(p)$ <u>be the kernel of the multiplication</u>

$$d^+ : M(p) \to M(p)$$

<u>where</u> $d(z) = d^+(z)d^-(z)$ <u>is a factorization of</u> $d(z)$ <u>with</u> $d^+(z)$ <u>having its roots in</u> $|z| < 1$, <u>while</u> $d^-(z)$ <u>has its roots in</u> $|z| > 1$. <u>Then the family of vector spaces</u> $M^+(p)$ <u>form a vector bundle over</u> $P(m,n)$ <u>with fibre dimension equal to the degree of</u> d^+ .

The passage from $M(p)$ to $M^+(p)$ is one of localizing to the interior of the unit circle. Thus if $f(z)$ is any polynomial with all roots in $|z| > 1$ and annihilating $M^-(p)$, then

$$M^+(p) \cong M(p)_f$$

where $M(p)_f = M(p) \otimes A_f$ and A_f is the ring of rational functions in z whose denominator is a power of f. This follows from the decomposition (2.3.2) (which holds for all $p \in P(m,n)$) since f acts as a unit on $M^+(p)$.

Using this observation we can now describe how M^+ behaves under products:

<u>LEMMA (2.3.6).</u> <u>Let</u> $q = pp'$ <u>with</u> $p \in P(m,n)$ <u>and</u> $p' \in P(m',n)$. <u>Then there is a natural exact sequence</u>

$$0 \to M^+(p') \to M^+(q) \to M^+(p) \to 0 .$$

<u>Proof</u>: There is an obvious exact sequence

$$0 \to M(p') \xrightarrow{p} M(q) \to M(p) \to 0 .$$

Now localize with respect to $f = d_q^- = d_p^- d_{p'}^-$. Since localization preserves exactness the lemma follows.

The maps in (2.3.6) are clearly continuous in p and p' so that we can interpret (2.3.6) as giving an exact sequence of vector bundles over $P(m,n) \times P(m',n)$. In particular taking $p' = z$, and noting that $M^+(p')$ is then the trivial bundle of rank n, we deduce

<u>COROLLARY (2.3.7).</u> <u>There is a natural exact sequence</u>

$$0 \to C^n \to M^+(zp) \to M^+(p) \to 0 .$$

We are now ready to apply these algebraic preliminaries to construct the homomorphism

$$\alpha : K^{-2}(X) \to K(X)$$

which will be the required inverse of the Bott periodicity map β.

From (2.1.9) we have the exact sequence

$$0 \to K^{-2}(X) \to K(S^2 \times X) \to K(X) \to 0$$

and we shall in fact define

$$\alpha : K(S^2 \times X) \to K(X)$$

before restricting it to the subgroup $K^{-2}(X)$.

Let S^2 be decomposed into the two balls B_0, B_∞ the inside and outside of the unit circle. Since these balls are contractible any vector bundle on $B_0 \times X$ or $B_\infty \times X$ is the pull-back of a bundle on X. Hence any vector bundle E on $S^2 \times X$ can be constructed from a vector bundle F on X, together with a continuous family of isomorphisms

$$f(z,x) : F_x \to F_x \qquad x \in X , \ |z| = 1 .$$

Moreover the isomorphism class of E depends only on the homotopy class of f. This is essentially just Lemma (1.4.9) with X replaced by $S^1 \times X$.

Now approximate f, uniformly in x, by a finite Fourier series

$$f_m(z,x) = \sum_{-m}^{m} a_k(x) z^k$$

where each a_k is a continuous endomorphism of F. This may be done for example by using the Cesaro means of the Fourier coefficients. Thus

$$\|f(z,x) - f_m(z,x)\| < \varepsilon \quad \text{for} \quad x \in X, \ |z| = 1 ,$$

where $\| \ \|$ denotes a matrix norm relative to some fixed metric on F. For small ε the endomorphism $f_m(z,x)$ will be invertible for all $x \in X$ and $|z| = 1$. Moreover, two different approximations f_m and f'_m will be connected by the linear segment $tf_m + (1-t)f'_m$ $(0 \le t \le 1)$ of isomorphisms.

Next we put

$$p(z,x) = z^m f_m(z,x)$$

so that p is a polynomial in z with coefficients in $\mathrm{End}(F)$. Moreover p is invertible on $|z| = 1$ for all x. We can therefore apply our algebraic construction and form the vector bundle $M^+(p)$ over X. Finally we define

$$\alpha[E] = [M^+(p)] - mn \in K(X) .$$

This is independent of the degree m in view of (2.2.7), and it is independent of the choice of f in its homotopy class because $K(X)$ is a homotopy invariant. Since α is clearly compatible with direct sums it induces a group homomorophism

$$\alpha : K(S^2 \times X) \to K(X) .$$

The basic properties of α are described in the following Proposition.

PROPOSITION (2.3.8). The homomorphism

$$\alpha : K(S^2 \times X) \to K(X)$$

has the properties

(1) it is functorial in X

(2) it is a $K(X)$-module homomorphism

(3) for X = point and $b \in K(S^2)$ the Bott generator, $\alpha(b) = 1$.

Proof: (1) is clear from the construction. For (2) we just have to note that, if 1_V denotes the identity automorphism of a vector space V, we have a natural isomorphism

$$M^+(p \otimes 1_V) \cong M^+(p) \otimes V .$$

Finally for (3) we recall that $b = H - 1$ and that H is constructed with $f = p = z$. Since $M^+(z) \cong C$, while $M^+(1) = 0$ it follows that $\alpha(b) = 1$.

Since the Bott map

$$\beta : K(X) \to K^{-2}(X)$$

is given by multiplying with b we see that (2.3.8) has the following immediate Corollary.

COROLLARY (2.3.9). $\alpha\beta = 1$, the identity of $K(X)$.

Note: It is immaterial here whether we regard α as defined on $K(S^2 \times X)$ or on its subgroup $K^{-2}(X)$.

To complete the proof of the periodicity theory it remains to prove that $\beta\alpha = 1$ on $K^{-2}(X)$. For this it will be convenient to extend β to a homomorphism

$$K^{-n}(X) \to K^{-n-2}(X) \qquad n \geq 0 .$$

We can similarly extend α to a homomorphism

$$(2.3.10) \qquad K^{-n-2}(X) \to K^{-n}(X) .$$

To do this we simply replace X by $S^n \times X$, and recall that

$$K^{-n}(X) = \mathrm{Ker}\{K(S^n \times X) \to K(X)\} .$$

The naturality of α then shows that it induces the required homomorphism (2.3.10) .

With α and β extended in this way we now prove

<u>LEMMA (2.3.11)</u>. $\alpha\beta = \beta\alpha$ <u>on</u> $K^{-2}(\;)$, i.e. <u>the diagram</u>

$$\begin{array}{ccc} K^{-2}(X) & \xrightarrow{\beta} & K^{-4}(X) \\ \downarrow\alpha & & \downarrow\alpha \\ K(X) & \xrightarrow{\beta} & K^{-2}(X) \end{array}$$

<u>commutes</u> .

<u>Proof</u>: Consider the map

$$\varepsilon : R^2 \times R^2 \to R^2 \times R^2$$

given by interchanging the two factors. This extends to a map of $S^4 = R^4 \cup \infty$ and so induces an automorphism ε^* of $K^{-4}(X)$. Since (13)(24) is an even permutation ε is in the identity component of $GL(4,R)$ and hence the corresponding map of S^4 is homotopic to the identity. Thus $\varepsilon^* = 1$. But the two ways of going round the square in (2.3.11) differ by ε^*. More precisely

$$\alpha\varepsilon^*\beta = \beta\alpha .$$

Since $\varepsilon^* = 1$ we get $\alpha\beta = \beta\alpha$ as required .

Remark: A general permutation σ of the factors of R^n induces a map σ^* on $K^{-n}(X)$. This depends only on the component of $GL(n,R)$ containing σ, i.e. on sign(σ). For the even case $\sigma^* = 1$ while for the odd case $\sigma^* = -1$ as follows from (2.1.6).

Lemma (2.3.11) and (2.3.9) together complete the proof of the periodicity theorem (2.3.1). As we have just seen everything follows formally once we have constructed α with the properties (1), (2), (3) in Proposition (2.3.8). There are a number of different ways to make this construction. We have given the most elementary algebraic method, but the use of Toeplitz operators as in [4] or of elliptic differential operators as in [2] provide alternative proofs. For certain generalizations, as explained in [2], these analytic methods are essential.

When applied with X = point the periodicity theorem implies

COROLLARY (2.3.12). $K^{-2}(\text{point}) = K^2(S^2) \cong Z$ *with* b *as generator*.

This Corollary amounts to the statement that

$$\pi_1(U(1)) \to \varinjlim \pi_1(U(n))$$

is an isomorphism. In fact it is easy to show by using the fibrations

$$U(n-1) \to U(n) \to S^{2n-1}$$

that

$$\pi_1(U(1)) \cong \pi_1(U(n)) \qquad \text{for all} \quad n \geq 1 .$$

In view of this the periodicity theorem is essentially equivalent to the following version:

THEOREM (2.3.13). *The tensor product of bundles induces an isomorphism*

$$K(S^2) \otimes K(X) \to K(S^2 \times X) .$$

Our proof of Theorem (2.3.13) generalizes, with only minor changes, to the case when $S^2 \times X$ is replaced by a fibration over X with fibre S^2 (and structure group $U(1)$). The essential point is that our construction of the homomorphism α is local over X. The variable z is not globally defined but any two local choices differ by $z \to uz$ with $|u| = 1$ (depending on x but not on z). We can formulate this generalization of (2.3.13) as follows. For any complex vector bundle E over X we can form the associated bundle $P(E)$ of projective spaces. In particular let

$$E = L \oplus 1$$

be the sum of a line-bundle L and the trivial line-bundle 1. Then $P(E)$ has fibre the complex projective line on S^2. Moreover on $P(E)$ we have the natural tautologous line-bundle H^*. Consideration of determinants then shows that, on $P(E)$, we have an exact sequence

$$0 \to H^* \to p^*(E) \to H \otimes p^*L \to 0$$

where $p : P(E) \to X$ denotes the projection. This exact sequence shows that, in the $K(X)$-module $K(P(E))$, the element $[H]$ satisfies the relation

$$[E] = [H][L] + [H]^*$$

or, since $E = L \oplus 1$,

$$([H] - 1)([H][L] - 1) = 0$$

The generalization of (2.3.13) can then be stated as follows:

THEOREM (2.3.14). Let L be a line-bundle over X. Then, as a $K(X)$-algebra, $K(P(L \oplus 1))$ is generated by $[H]$ and is subject to the single relation $([H] - [1])([H][L] - [1]) = 0$.

We have already noted the quadratic relation satisfied by $[H]$. The essential point is that as a $K(X)$-module $K(P(E))$ is free on the 2 generators $[1]$, $[H]$. Note that the Bott generator corresponds to $[H][L] - [1]$.

In the proof of the main periodicity theorem (2.3.1) we have already introduced the higher K-groups $K^{-n}(X)$. The periodicity theorem naturally extends to these:

THEOREM 2.3.15. For any space X and any $n \leq 0$, the map $K^{-2}(\text{point}) \otimes K^{-n}(X) \to K^{-n-2}(X)$ induces an isomorphism $\beta : K^{-n}(X) \to K^{-n-2}(X)$.

Proof: $K^{-2}(\text{point}) = \widetilde{K}(S^2)$ is the free abelian group generated by $[H] - [1]$. If $(X, Y) \in C^2$ the maps in the exact sequence (2.4.4) all commute with the periodicity isomorphism β. This is immediate for i^* and j^* and is also true for δ since this was also induced by a map of spaces. In other words β shifts the whole sequence to the left by six terms. Hence if we define $K^n(X, Y)$ for $n > 0$ inductively by $K^{-n} = K^{-n-2}$ we can extend (2.2.5) to an exact sequence infinite in both directions. Alternatively using the periodicity β we can define an exact sequence of six terms

$$\begin{array}{ccccc} K^0(X,Y) & \longrightarrow & K^0(X) & \longrightarrow & K^0(Y) \\ \uparrow & & & & \downarrow \\ K^1(Y) & \longleftarrow & K^1(X) & \longleftarrow & K^1(X,Y) \end{array} .$$

Except when otherwise stated we shall now always identify K^n and K^{n-2}. We introduce

$$K^*(X) = K^0(X) \oplus K^1(X) .$$

We define $K^*(X)$ to be $K^0(X) \oplus K^1(X)$. Then, for any pair (X, Y), we have an exact sequence

$$\begin{array}{ccccc} K^0(X,Y) & \longrightarrow & K^0(X) & \longrightarrow & K^0(Y) \\ \uparrow & & & & \downarrow \\ K^1(Y) & \longleftarrow & K^1(X) & \longleftarrow & K^1(X,Y) \end{array} .$$

§ 2.4. $K_G(X)$. Suppose that G is a finite group and that X is a G-space. Let $\mathrm{Vect}_G(X)$ denote the set of isomorphism classes of G-vector bundles over X. This is an abelian semi-group under $\oplus$. We form the associated abelian group and denote it by $K_G(X)$. If $G = 1$ is the trivial group then $K_G(X) = K(X)$. If on the other hand X is a point then $K_G(X) \cong R(G)$ the character ring of G.

If E is a G-vector bundle over X then $P(E)$ is a G-space. If $E = L \oplus 1$ when L is a G-bundle then the zero and infinite sections $X \to P(E)$ are both G-sections. Also the bundle H over $P(E)$ is a G- line bundle. If we now examine the proof of the periodicity theorem which we have just given we see that we could have assumed a G-action on everything. Thus we get the periodicity theorem for K_G:

THEOREM 2.4.1. *If X is a G-space, and if L is a G-line bundle over X, the map $t \to [H]$ induces an isomorphism of $K_G(X)$ - modules:*

$$K_G(X)[t]/(t[L] - 1)(t - 1) \longrightarrow K_G(P(L \oplus 1)) \quad .$$

§2.5. Computations of $K^*(X)$ for some X.

From the periodicity theorem, we see that $\widetilde{K}(S^n) = 0$ if n is odd, and $\widetilde{K}(S^n) = Z$ if n is even. This allows us to prove the Brouwer fixed point theorem.

THEOREM 2.5.1. Let D^n be the unit disc in Euclidean n-space. If $f : D^n \to D^n$ is continuous, then for some $x \in D^n$, $f(x) = x$.

Proof: Since $\widetilde{K}^*(D^n) = 0$, and $\widetilde{K}^*(S^{n-1}) \neq 0$, S^{n-1} is not a retract of D^n. If $f(x) \neq x$ for every $x \in D^n$, define $g : D^n \to S^{n-1}$ by $g(x) = (1 - \alpha(x))f(x) + \alpha(x)x$, where $\alpha(x)$ is the unique function such that $\alpha(x) \geq 0$, $|g(x)| = 1$. If $f(x) \neq x$ for all x, clearly such a function $\alpha(x)$ exists. If $x \in S^{n-1}$, $\alpha(x) = 1$, and $g(x) = x$. Thus g is a retraction of D^n onto S^{n-1}.

We will say that a space X is a cell complex if there is a filtration by closed sets $X_{-1} \subset X_0 \subset X_1 \subset \cdots \subset X_n = X$ such that each $X_k - X_{k-1}$ is a disjoint union of open k-cells, and $X_{-1} = \emptyset$.

PROPOSITION 2.5.2. If X is a cell complex such that $X_{2k} = X_{2k+1}$ for all k,

$K^1(X) = 0$

$K^0(X)$ is a free abelian group with generators in a one-one correspondence with the cells of X.

Proof: We proceed by induction on n. Since X_{2n}/X_{2n-2} is a union of 2n-spheres with a point in common we have:

$$K^1(X_{2n}, X_{2n-2}) = 0$$

$$K^0(X_{2n}, X_{2n-2}) = \mathbb{Z}^k$$

where k is the number of 2n-cells in X. The result for X_{2n} now follows from the inductive hypothesis and the exact sequence of the pair (X_{2n}, X_{2n-2}). As examples of spaces to which this proposition applies, we may take X to be a complex Grassmann manifold, a flag manifold, a complex quadric (a space whose homogeneous defining equation is of the form $\Sigma z_i^2 = 0$). We shall return to the Grassmann and flag manifolds in more detail later.

PROPOSITION 2.5.3. Let $L_1, \cdots, L_n$ be line bundles over X, and let H be the standard bundle over $P(L_1 \oplus \cdots \oplus L_n)$. Then, the map $t \to [H]$ induces an isomorphism of K(X)-modules

$$K(X)[t] \Big/ \prod_{i=1}^{n} (t - [L_i^*]) \longrightarrow K(P(L_1 \oplus \cdots \oplus L_n)) .$$

Proof: First we shall show that we may take $L_n = 1$. In fact for any vector bundle E and line bundle L over X we have $P(E \otimes L) = P(E)$ and the standard line bundles G, H over $P(E \otimes L^*)$, $P(E)$ are related by $G^* = H^* \otimes L$, i.e., $G = H \otimes L^*$. Taking $E = L_1 \oplus \cdots \oplus L_n$ and $L = L_n^*$ we see that the propositions for $L_1 \oplus \cdots \oplus L_n$ and for $M_1 \oplus \cdots \oplus M_n$ with $M_i = L_i \otimes L_n^*$ are equivalent. We shall suppose therefore that $L_n = 1$ and for brevity write

$$P_m = P(L_1 \oplus \cdots \oplus L_m) \qquad \text{for } 1 \leq m \leq n$$

so that we have inclusions $X = P_1 \to P_2 \to \cdots \to P_n$. If H_m denotes the standard line bundle over P_m then $H_m | P_{m-1} \cong H_{m-1}$. Now we observe that we have a commutative diagram

$$\begin{array}{ccc} P_{n-1} & \xrightarrow{\ s\ } & P(H^*_{n-1} \oplus 1) \\ \Big\downarrow{\scriptstyle \pi_{n-1}} & & \Big\downarrow{\scriptstyle q} \\ P_1 & \xrightarrow{\ i_n\ } & P_n \end{array}$$

(π_{n-1} is the projection onto $X = P_1$, i_n is the inclusion, s is the zero section) which induces a homeomorphism

$$P(H^*_{n-1} \oplus 1)/s(P_{n-1}) \longrightarrow P_n/P_1 \quad .$$

Moreover $q^*(H_n) \cong G$, the standard line bundle over $P(H^*_{n-1} \oplus 1)$. Now $K(P(H^*_{n-1} \oplus 1))$ is a free $K(P_{n-1})$-module on two generators

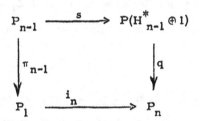

[1] and [G], and [G] satisfies the equation $([G] - [1])([G] - [H_{n-1}]) = 0$. Since $s^*[G] = [1]$ it follows that $K(P(H^*_{n-1} \otimes 1), s(P_{n-1}))$ is the submodule generated freely by $[G] - [1]$ and that, on this submodule, multiplication by $[G]$ and $[H_{n-1}]$ coincide. Hence $K(P_n, P_1)$ is a free $K(P_{n-1})$-module generated freely by $([H_n] - [1])$ and this module structure is such that, for any $x \in K(P_n, P_1)$,

$$[H_{n-1}]x = [H_n]x .$$

Now assume the proposition established for $n - 1$, so that

$$K(P_{n-1}) \cong K(X)[t]\Big/\prod_{i=1}^{n-1}(t - [L_i^*])$$

with t corresponding to $[H_{n-1}]$. Then it follows that $t \to [H_n]$ induces an isomorphism of the ideal $(t - 1)$ in

$$K(X)[t]\Big/(t - 1)\prod_{i=1}^{n-1}(t - [L_i^*])$$

onto $K(P_n, P_1)$. Since

$$K(P_n) \cong K(P_n, P_1) \oplus K(X)$$

and since $L_n = 1$ this gives the required result for $K(P_n)$ establishing the induction and completing the proof.

COROLLARY 2.5.4. $K(P(C^n)) \cong \mathbb{Z}[t]/(t-1)^n$ <u>under the map</u> $t \to [H]$.

<u>Proof:</u> Take X to be a point.

We could again have assumed that a finite group acted on everything, and we would have obtained

$$K_G(X)[t] \Big/ \prod_{i=1}^{n} (t - [L_i^*]) \cong K_G(P(L_1 \oplus \cdots \oplus L_n)) \quad .$$

§2.6. Multiplication in $K^*(X, Y)$. We first observe that the multiplication in $K(X)$ can be defined "externally" as follows. Let E, F be two bundles over X, and let $E \hat{\otimes} F$ be $\pi_1^*(E) \otimes \pi_2^*(F)$ over $X \times X$. If $\Delta : X \to X \times X$ is the diagonal, then $E \otimes F = \Delta^*(E \hat{\otimes} F)$.

If E is a bundle on X, F a bundle on Y, let $E \hat{\otimes} F = \pi_X^*(E) \otimes \pi_Y^*(F)$ on $X \times Y$. This defines a pairing

$$K(X) \otimes K(Y) \longrightarrow K(X \times Y) \quad .$$

If X, Y have basepoints, $\widetilde{K}(X \wedge Y)$ is the kernel of $\widetilde{K}(X \times Y) \longrightarrow \widetilde{K}(X) \oplus \widetilde{K}(Y)$. Thus, we have $\widetilde{K}(X) \otimes \widetilde{K}(Y) \to \widetilde{K}(X \wedge Y)$.

Suppose that (X, A), (Y, B) are pairs. Then we have

$$\widetilde{K}(X/A) \otimes \widetilde{K}(Y/B) \longrightarrow \widetilde{K}((X/A) \wedge (Y/B)) \quad .$$

That is,

$$K(X,A) \otimes K(Y,B) \longrightarrow K(X \times Y, (X \times B) \cup (A \times Y)) \, .$$

We define $(X,A) \times (Y,B)$ to be $(X \times Y, (X \times B) \cup (A \times Y))$.

In the special case that $X = Y$, we have a diagonal map $\Delta : (X, A \cup B) \to (X, A) \times (X, B)$. This gives us $K(X,A) \otimes K(X,B) \longrightarrow K(X, A \cup B)$. In particular, taking $B = \emptyset$, we see that $K(X,A)$ is a $K(X)$-module. Further, it is easy to see that

$$K(X,A) \longrightarrow K(X) \longrightarrow K(A)$$

is an exact sequence of $K(X)$-modules.

More generally, we can define products

$$K^{-n}(X,A) \otimes K^{-m}(Y,B) \longrightarrow K^{-n-m}((X,A) \times (Y,B))$$

for $m, n \leq 0$ as follows:

$$K^{-n}(X,A) = \widetilde{K}(S^n \wedge (X/A))$$

$$K^{-m}(Y,B) = \widetilde{K}(S^m \wedge (Y/B)) \quad .$$

Thus, we have

$$\begin{aligned} K^{-n}(X,A) \otimes K^{-m}(Y,B) &\longrightarrow \widetilde{K}(S^n \wedge (X/A) \wedge S^m \wedge (Y/B)) \\ &= \widetilde{K}(S^n \wedge S^m \wedge (X/A) \wedge (Y/B)) \\ &= K^{-n-m}((X,A) \times (Y,B)) \quad . \end{aligned}$$

Thus, if we define $xy \in K^{-n-m}(X, A \cup B)$ for $x \in K^{-n}(X,A)$, $y \in K^{-m}(X,B)$ to be $\Delta^*(x \otimes y)$, where $\Delta : (X, A \cup B) \to (X,A) \times (X,B)$ is the diagonal, then (2.4.11) shows that $xy = (-1)^{mn} yx$.

We define $K^{\#}(X,A)$ to be

$$\sum_{n=0}^{\infty} K^{-n}(X, A) \quad .$$

Then $K^{\#}(X)$ is a graded ring, and $K^{\#}(X, A)$ is a graded $K^{\#}(X)$-module. If $\beta \in K^{-2}(\text{point})$ is the generator, multiplication by β induces an isomorphism $K^{-n}(X, A) \to K^{-n-2}(X, A)$ for all n. We define $K^{*}(X, A)$ to be $K^{\#}(X, A)/(1 - \beta)$. Then $K^{*}(X)$ is a ring graded by $\mathbb{Z}_2$, and $K^{*}(X, A)$ is a $\mathbb{Z}_2$-graded module over $K^{*}(X)$.

For any pair (X, A), each of the maps in the exact triangle

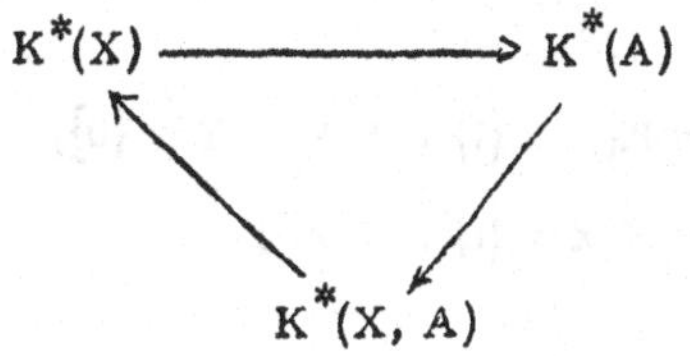

is a $K^{*}(X)$-module map. Only the coboundary δ causes any difficulty and so we need to prove

LEMMA 2.6.0. $\delta : K^{-1}(Y) \to K^{0}(X, Y)$ is a $K(X)$-module homomorphism.

Proof: By definition δ is induced by the inclusion of pairs $j : (X \times \{1\} \cup Y \times I,\ Y \times \{0\}) \to (X \times \{1\} \cup Y \times I,\ Y \times \{0\} \cup X \times \{1\})$ (see figure)

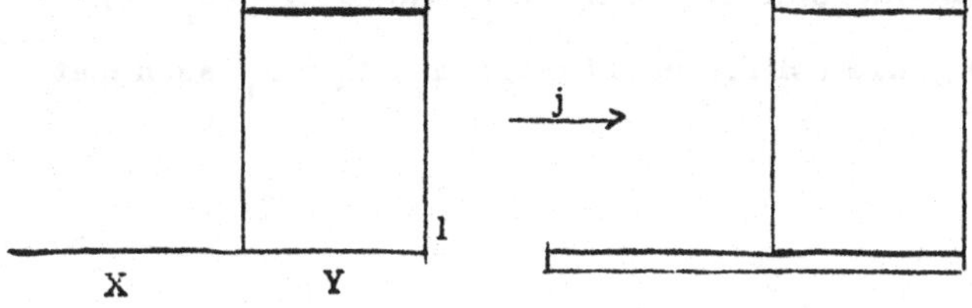

Hence $\delta = j^*$ is a module homomorphism over the absolute group

$$K(X \times \{1\} \cup Y \times I) \cong K(X) \quad .$$

It remains only to observe that the $K(X)$-module structures of the two groups involved are the standard ones. For $K^{-1}(Y)$ this is immediate and for $K(X, Y)$ we have only to observe that the projection $I \to \{1\}$ induces the isomorphisms

$$K(X, Y) \to K(X \times \{1\} \cup Y \times I, \ Y \times \{0\})$$
$$K(X) \to K(X \times \{1\} \cup Y \times I) \quad .$$

We shall now digress for some time to give an alternative and often illuminating description of $K(X, A)$ which has particular relevance for products.

If $n \geq 1$, we define $C_n(X, A)$ to be a category as follows: An object of $C_n(X, A)$ is a collection $E_n, E_{n-1}, \cdots, E_0$ of bundles over X, together with maps $\alpha_i : E_i|A \to E_{i-1}|A$ such that

$$0 \longrightarrow E_n|A \xrightarrow{\alpha_n} E_{n-1}|A \cdots \xrightarrow{\alpha_1} E_0|A \longrightarrow 0$$

is exact. The morphisms $\varphi : E \to F$, where $E = (E_i, \alpha_i)$ $F = (F_i, \beta_i)$, are collections of maps $\varphi_i : E_i \to F_i$ such that

$\beta_i \varphi_i = \varphi_{i-1} \alpha_i$. In particular, $C_1(X,A)$ consists of pairs of bundles E_1, E_0 over X and isomorphisms $\alpha : E_1|A \cong E_0|A$.

An elementary sequence in $C_n(X,A)$ is a sequence of the form $0, 0, \cdots, 0, E_p, E_{p-1}, 0, \cdots, 0$ where $E_p = E_{p-1}$, α = identity map. We define $E \sim F$ if for some set of elementary objects $Q_1, \cdots, Q_n, P_1, \cdots, P_m$,

$$E \oplus Q_1 \oplus \cdots \oplus Q_n \cong F \oplus P_1 \oplus \cdots \oplus P_m \quad .$$

The set of such equivalence classes is denoted by $\mathcal{L}_n(X,A)$. It is clear that $\mathcal{L}_n(X,A)$ is a semigroup for each n.

There is a natural inclusion $C_n(X,A) \subset C_{n+1}(X,A)$ which induces a homomorphism $\mathcal{L}_n(X,A) \to \mathcal{L}_{n+1}(X,A)$. We denote by $C_\infty(X,A)$ the union of all of the $C_n(X,A)$, and by $\mathcal{L}_\infty(X,A)$ the direct limit of the $\mathcal{L}_n(X,A)$.

The main theorem of this section is the following:

THEOREM 2.6.1. For all $n \geq 1$, the maps $\mathcal{L}_n(X,A) \to \mathcal{L}_{n+1}(X,A)$ are isomorphisms, and $\mathcal{L}_n(X, A) \cong K(X,A)$.

We shall break up the proof of this theorem into a number of lemmas.

Consider first the special case $A = \emptyset$, $n = 1$. Then $C_1(X, \emptyset)$ consists of all pairs E_1, E_0 of bundles. We see that $(E_1, E_0) \sim (F_1, F_0)$ if and only if there are bundles Q, P such that $E_1 \oplus Q \cong F_1 \oplus P$, $E_0 \oplus Q \cong F_0 \oplus P$. Then the map $\mathcal{L}_1(X, \phi) \to K(X)$ given by $(E_1, E_0) \to [E_0] - [E_1]$ is an isomorphism. In fact $\mathcal{L}_1(X, \phi)$ coincides with one of our definitions of $K(X)$.

DEFINITION 2.6.2. An Euler characteristic χ_n for $\mathcal{L}_n$ is a transformation of functors

$$\chi_n : \mathcal{L}_n(X, A) \longrightarrow K(X, A)$$

such that whenever $A = \phi$, $\chi(E_n, E_{n-1}, \cdots, E_0) = \Sigma (-1)^i [E_i]$.

To begin we need a simple lemma.

LEMMA 2.6.3. Let $A \subset X$, and let E, F be bundles over X. Let $\varphi : E|A \to F|A$, $\psi : E \to F$ be monomorphisms (resp. isomorphisms) and assume $\psi|A$ is homotopic to φ. Then φ extends to X as a monomorphism (resp. isomorphism).

Proof: Let $Y = (A \times [0, 1]) \cup (X \times [0])$. Then, if E', F' are the inverse images of E, F under the projection $Y \to X$, we can define $\Phi : E' \to F'$ which is a monomorphism (resp. isomorphism) such that $\Phi|A \times [1] = \varphi$, $\Phi|X \times [0] = \psi$. We can extend Φ to $(U \times [0, 1]) \cup (X \times [0])$ for some neighborhood U of A. Let $f : X \to [0, 1]$ be such that $f(A) = 1$, $f(X - U) = 0$.

Let $\varphi_x = \Phi_{(x, f(x))}$. Then this extends φ to X.

LEMMA 2.6.4. *If* A *is a point,*

$$0 \longrightarrow \mathcal{L}_1(X, A) \longrightarrow \mathcal{L}_1(X) \longrightarrow \mathcal{L}_1(A)$$

is exact. Thus, if χ_1 *is an Euler characteristic for* $\mathcal{L}_1$, $\chi_1 : \mathcal{L}_1(X, A) \to K(X, A)$ *is an isomorphism when* A *is a point.*

Proof: If (E_1, E_0) represents an element of $\mathcal{L}_1(X)$ whose image in $\mathcal{L}_1(A)$ is zero, E_1 and E_0 have the same dimension over A. Thus there is an isomorphism $\varphi: E_1|A \to E_0|A$. Thus we have exactness for $\mathcal{L}_1(X, A) \to \mathcal{L}_1(X) \to \mathcal{L}_1(A)$.

If (E_1, E_0, φ) has image zero in $\mathcal{L}_1(X)$, there is a trivial P and an isomorphism $\psi : E_1 \oplus P \cong E_0 \oplus P$. $\psi(\varphi \oplus 1)^{-1}$ is an automorphism of $E_0 \oplus P|A$. Since A is a point any such automorphism must be homotopic to the identity and hence by (2.6.3) it extends to $\alpha : E_0 \oplus P \cong E_0 \oplus P$. Thus, we have a commuting diagram:

$$\begin{array}{ccc}
(E_1 \oplus P)|A & \xrightarrow{\varphi \oplus 1} & (E_0 \oplus P)|A \\
\Big\downarrow \psi|A & & \Big\downarrow \alpha|A \\
(E_0 \oplus P)|A & \xrightarrow{1} & (E_0 \oplus P)|A
\end{array} \quad .$$

Thus (E_1, E_0, φ) represents 0 in $\mathcal{L}_1(X, A)$. Thus $\mathcal{L}_1(X,A) \longrightarrow \mathcal{L}_1(X)$ is an injection.

LEMMA 2.6.5. $\mathcal{L}_1(X/A, A/A) \to \mathcal{L}_1(X,A)$ *is an isomorphism for all* (X,A). *Thus, if* χ_1 *is an Euler characteristic,* $\chi_1 : \mathcal{L}_1(X,A) \to K(X,A)$ *is an isomorphism for all* (X, A).

Proof: Since the isomorphism $\mathcal{L}_1(X/A, A/A) \to K(X,A)$ factors through $\mathcal{L}_1(X,A)$, the map $\mathcal{L}_1(X/A, A/A) \to \mathcal{L}_1(X,A)$ is injective.

Suppose that E_1, E_0 are bundles on X, $\alpha : E_1|A \to E_0|A$ is an isomorphism. Let P be a bundle on X such that there is an isomorphism $\beta : E_1 \oplus P \to F$, where F is trivial. Then (E_1, E_0, α) is equivalent to $(F, E_0 \oplus P, \gamma)$ where $\gamma = (\alpha \oplus 1)\beta^{-1}$. Then, $(F, E_0 \oplus P, \gamma)$ is the image of $(F, (E_0 \oplus P)/\gamma, \gamma/\gamma)$. Thus, $\mathcal{L}_1(X/A, A/A) \to \mathcal{L}_1(X,A)$ is onto.

LEMMA 2.6.6. *If* χ_1, χ_1' *are two Euler characteristics for* $\mathcal{L}_1$, $\chi_1 = \chi_1'$.

Proof: $\chi_1' \chi_1^{-1}$ is a transformation of functors from K to itself which is the identity on each $K(X)$. Since $K(X,A) = \widetilde{K}(X/A)$ is injected into $K(X/A)$, it is the identity on all $K(X,A)$.

LEMMA 2.6.7. *There exists an Euler characteristic* χ_1 *for* $\mathcal{L}_1$.

Proof: Suppose (E_1, E_0, α) represents an element of $\mathcal{L}_1(X,A)$. Let X_0, X_1 be two copies of X, and let $Y = X_0 \cup_A X_1$ be the space which results from identifying corresponding points of A . Then $[E_1, \alpha, E_0] \in K(Y)$. Let $\pi_i : Y \to X_i$ be the obvious retraction. Then $K(Y) = K(Y, X_i) \oplus K(X_i)$. The map $(X_0, A) \longrightarrow (Y, X_1)$ induces an isomorphism $K(Y, X_1) \to K(X_0, A)$. Let $\chi_1(E_1, E_0, \alpha)$ be the image of the component of $[E_1, \alpha, E_0]$ which lies in $K(Y, X_1)$. If $A = \emptyset$, then $\chi(E_1, E_0, \alpha) = [E_0] - [E_1]$. One can easily verify that this definition is independent of the choices made.

COROLLARY 2.6.8. *The class of* (E_1, E_0, α) *in* $\mathcal{L}_1(X, A)$ *only depends on the homotopy class of* α .

Proof: Let $Y = X \times [0, 1]$, $B = A \times [0, 1]$. Then, if α_t is a homotopy with $\alpha_0 = \alpha$, α_t defines $\beta : \pi^*(E_1)|B \cong \pi^*(E_0)|B$. Let $i_j : (X,A) \to (X \times [j], A \times [j])$. From the commuting diagram

$$\begin{array}{ccccc} \mathcal{L}_1(X,A) & \xleftarrow{\ i_0^*\ } & \mathcal{L}_1(Y,B) & \xrightarrow{\ i_1^*\ } & \mathcal{L}_1(X,A) \\ \downarrow{\scriptstyle \chi_1} & & \downarrow{\scriptstyle \chi_1} & & \downarrow \\ K(X,A) & \xleftarrow{\ i_0^*\ } & K(Y,B) & \xrightarrow{\ i_1^*\ } & K(X,A) \end{array}$$

we see that since every map is an isomorphism, and since $i^*_0(i^*_1)^{-1}$ is the identity, $(E_1, E_0, \alpha_0) = (E_1, E_0, \alpha_1)$.

LEMMA 2.6.9. The map $\mathcal{L}_n(X,A) \to \mathcal{L}_{n+1}(X,A)$ is onto for $n \geq 1$.

Proof: If $(E_{n+1}, \cdots, E_0; \alpha_{n+1}, \cdots, \alpha_1)$ represents an element of $\mathcal{L}_{n+1}(X,A)$, so does

$$(E_{n+1}, E_n \oplus E_{n+1}, E_{n-1} \oplus E_{n+1}, E_{n-2}, \cdots, E_0; \alpha_{n+1}, \alpha_n \oplus 1, \cdots, \alpha_1) .$$

The two maps $\alpha_{n+1} \oplus 0 : E_{n+1} \to E_n \oplus E_{n+1}$ and $0 \oplus 1 : E_{n+1} \to E_n \oplus E_{n+1}$ are (linearly) homotopic as monomorphisms. $0 \oplus 1$ extends to X, and thus by (2.6.3) $\alpha_{n+1} \oplus 0$ extends to a monomorphism $\beta : E_{n+1} \to E_n \oplus E_{n+1}$ on all of X. Thus we can write $E_n \oplus E_{n+1}$ as $\beta(E_{n+1}) \oplus Q$. Then we see that, if $\gamma : Q \to E_{n-1} \oplus E_{n+1}$ is the resulting map, $(E_{n+1}, \cdots, E_0; \alpha_{n+1}, \cdots, \alpha_1)$ is equivalent to $(0, Q, E_{n-1} \oplus E_{n+1}, \cdots, E_0; 0, \gamma, \cdots, \alpha_1)$. Thus $\mathcal{L}_n(X,A) \longrightarrow \mathcal{L}_{n+1}(X,A)$ is onto.

LEMMA 2.6.10. The map $\mathcal{L}_n(X,A) \to \mathcal{L}_{n+1}(X,A)$ is an isomorphism for all $n \geq 1$.

Proof: It suffices to produce a map $\mathcal{L}_{n+1}(X,A) \to \mathcal{L}_1(X,A)$ which is a left inverse of the map $\mathcal{L}_1(X,A) \to \mathcal{L}_{n+1}(X,A)$.

Let $(E_n, \cdots, E_0; \alpha_n, \cdots, \alpha_1)$ represent an element of $\mathcal{L}_n(X, A)$. Choose a Hermitian metric on each E_i. Let $\alpha_i' : E_{i-1}|A \to E_i|A$ be the Hermitian adjoint of α_i.

Put $F_0 = \Sigma E_{2i}$, $F_1 = \Sigma E_{2i+1}$, and define $\beta : F_1 \to F_0$ by $\beta = \Sigma \alpha_{2i+1} + \Sigma \alpha_{2i}'$. Then $(F_1, F_0, \beta) \in \mathcal{L}_1(X, A)$. This gives us a map $\mathcal{L}_n(X,A) \to \mathcal{L}_1(X,A)$. To see that it is well defined, we need only see that it does not depend on the choice of metrics. But all choices of metric are homotopic to one another, so that a change of metrics only changes the homotopy class of β. Thus this map is well defined. It clearly is a left inverse to $\mathcal{L}_1(X,A) \to \mathcal{L}_n(X, A)$.

COROLLARY 2.6.11. <u>For each</u> n <u>there exists exactly one Euler characteristic</u> $\chi_n : \mathcal{L}_n(X, A) \to K(X,A)$, <u>and it is always an isomorphism. Thus, there exists</u> $\chi : \mathcal{L}_\infty(X,A) \to K(X,A)$ <u>isomorphically.</u>

We next want to construct pairings

$$\mathcal{L}_n(X, Y) \otimes \mathcal{L}_m(X', Y') \longrightarrow \mathcal{L}_{n+m}((X, Y) \times (X', Y'))$$

compatible with the pairings

$$K(X, Y) \otimes K(X', Y') \longrightarrow K((X, Y) \times (X', Y')) \quad .$$

To do this, we must consider complexes of vector bundles, i.e., sequences

$$0 \longrightarrow E_n \xrightarrow{\sigma_n} E_{n-1} \xrightarrow{\sigma_{n-1}} \cdots \longrightarrow E_0 \longrightarrow 0 \quad .$$

where $\sigma_i \sigma_{i+1} = 0$ for all i .

LEMMA 2.6.12. <u>Let</u> $E_0, \cdots, E_n$ <u>be vector bundles on</u> X , <u>and let</u> $\sigma_i : E_i | Y \to E_{i-1} | Y$ <u>be such that</u>

$$0 \longrightarrow E_n \xrightarrow{\sigma_n} E_{n-1} \xrightarrow{\sigma_{n-1}} \cdots E_0 \longrightarrow 0$$

<u>is exact on</u> Y . <u>Then the</u> σ_i <u>can be extended to</u> $\rho_i : E_i \to E_{i-1}$ <u>on</u> X <u>such that</u> $\rho_i \rho_{i+1} = 0$ <u>for all</u> i .

<u>Proof:</u> We shall show that there is some open neighborhood U of Y in X and an extension τ_i of σ_i to U for all i such that

$$0 \longrightarrow E_n \xrightarrow{\tau_n} E_{n-1} \xrightarrow{\tau_{n-1}} \cdots \longrightarrow E_0 \longrightarrow 0$$

is exact on U. The extension to the whole of X is then achieved by replacing τ_i by $\rho\,\tau_i$ where ρ is a continuous function on X such that $\rho = 1$ on Y and supp $\rho \subset U$.

Suppose that on some closed neighborhood U_i of Y in X, we could extend $\sigma_1, \cdots, \sigma_i$ to $\tau_1, \cdots, \tau_i$ such that on U_i,

$$E_i \xrightarrow{\tau_i} E_{i-1} \longrightarrow \cdots \longrightarrow E_0 \longrightarrow 0$$

is exact. Let K_i be the kernel of τ_i on U_i. Then σ_{i+1} defines a section of $\mathrm{Hom}(E_{i+1}, K_i)$ defined on Y. Thus, this section can be extended to a neighborhood of Y in U_i, and thus $\sigma_{i+1} : E_{i+1} \to K_i$ can be extended to $\tau_{i+1} : E_{i+1} \to K_i$ on this neighborhood. σ_{i+1} is a surjection on Y, so τ_{i+1} will be a surjection on some closed neighborhood U_{i+1} of Y in U_i. Thus, the lemma follows by induction on i.

We introduce the set $\mathcal{S}_n(X, Y)$ of complexes of length n on X which are acyclic (i.e., exact) on Y. We say that two such complexes are homotopic if they are isomorphic to the restrictions to $X \times \{0\}$ and to $X \times \{1\}$ of an element in $\mathcal{S}_n(X \times I, Y \times I)$. There is a natural map

$$\Phi : \mathcal{S}_n(X, Y) \longrightarrow \mathcal{L}_n(X, Y)$$

given by restriction of homomorphisms.

LEMMA 2.6.13. <u>Φ induces a bijection of homotopy classes.</u>

<u>Proof:</u> The last lemma shows that Φ is surjective. To show that Φ is injective we have to show that any complex over $X \times \{0\} \cup X \times \{1\} \cup Y \times I$ which is acyclic over $Y \times I$ can be extended to a complex on the whole of $X \times I$. We carry out this extension in three steps. First we make the obvious extensions to $X \times [0, 1/4]$ and $X \times [3/4, 1]$. Next we apply the preceding lemma to the pair $X \times [1/4, 3/4]$, $Y \times [1/4, 3/4] \cup V \times \{1/4\} \cup V \times \{3/4\}$ where V is a closed neighborhood of Y in X over which the given complexes are still acyclic. This gives a complex on $X \times [1/4, 3/4]$ which agrees with that already defined at the two thickened ends along the strips $V \times \{1/4\}$ and $V \times \{3/4\}$. Thus if we now multiply everything by a function ρ such that

(i) $\rho = 1$ on $X \times \{0\} \cup X \times \{1\} \cup Y \times I$

(ii) $\rho = 0$ on $(X - V) \times \{1/4\} \cup (X - V) \times \{3/4\}$,

we obtain the desired extension (*see figure: the dotted line indicates the support of* ρ).

I

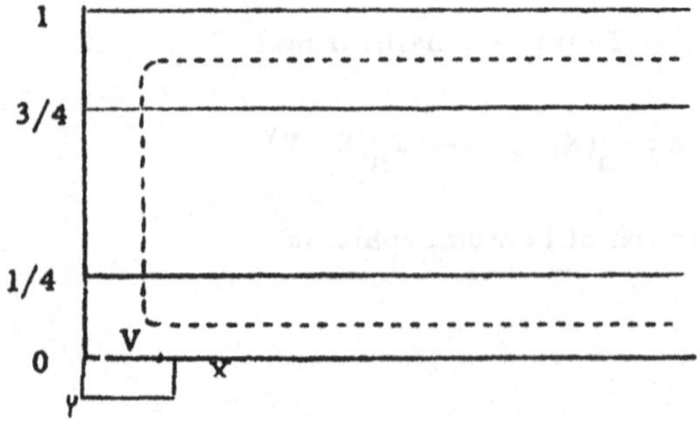

If $E \in \mathcal{E}_n(X, Y)$, $F \in \mathcal{E}_m(X', Y')$ then $E \otimes F$ is a complex on $X \times X'$ which is acyclic on $(X \times Y') \cup (Y \times X')$. Thus we have a natural pairing

$$\mathcal{E}_n(X, Y) \otimes \mathcal{E}_m(X', Y') \longrightarrow \mathcal{E}_{n+m}((X, Y) \times (X', Y'))$$

which is compatible with homotopies. Thus, by means of Φ, it induces a pairing

$$\mathcal{L}_n(X, Y) \otimes \mathcal{L}_m(X', Y') \longrightarrow \mathcal{L}_{n+m}((X, Y) \times (X', Y')) \quad .$$

LEMMA 2.6.14. <u>For any classes</u> $x \in \mathcal{L}_n(X, Y)$, $x' \in \mathcal{L}_m(X', Y')$,

$$\chi(x \otimes x') = \chi(x)\chi(x') \quad .$$

<u>Proof:</u> This is clearly true when $Y = Y' = \emptyset$. However, the pairing $K(X, Y) \otimes K(X', Y') \longrightarrow K((X, Y) \times (X', Y'))$ which we defined earlier was the only natural pairing compatible with the pairings defined for the case $Y = Y' = \emptyset$.

With this lemma we now have a very convenient description of the relative product. As a simple application we shall give a new construction for the generator of $\tilde{K}(S^{2n})$.

Let V be a complex vector space and consider the exterior algebra $\Lambda^*(V)$. We can regard this in a natural way as a complex of vector bundles over V. Thus we put $E_i = V \times \Lambda^i(V)$, and define

$$V \times \Lambda^i(V) \longrightarrow V \times \Lambda^{i+1}(V)$$

by

$$(v, w) \longrightarrow (v, v \wedge w) .$$

If $\dim V = 1$ the complex has just one map and this is an isomorphism for $v \neq 0$. Thus it defines an element of $K(B(V), S(V)) \cong \tilde{K}(S^2)$ where $B(V)$, $S(V)$ denote the unit ball and unit sphere of V with respect to some metric. Moreover this element is, from its definition, the canonical generator of $\tilde{K}(S^2)$ except for a sign -1. Since

$$\Lambda^*(V \oplus W) \cong \Lambda^*(V) \otimes \Lambda^*(W)$$

it follows that for any V, $\Lambda^*(V)$ defines a complex over V acyclic on $V - \{0\}$, and that this gives the canonical generator of $\tilde{K}(B(V), S(V)) = \tilde{K}(S^{2n})$ except for a factor $(-1)^n$ (where $n = \dim V$).

More generally the same construction applies to a vector bundle V over a space X. Let us introduce the Thom space X^V defined as the *one-point compactification* of V *or equivalently* as $B(V)/S(V)$. Then $K(B(V, S(V)) \cong \tilde{K}(X^V)$ and the exterior algebra of V defines an *element of* $\tilde{K}(X^V)$ which we denote by λ_V. It has the two properties

(A) λ_V restricts to a generator of $\tilde{K}(P^V)$ for each point $P \in X$.

(B) $\lambda_{V \oplus W} = \lambda_V \cdot \lambda_W$, where this product is from $\tilde{K}(X^V) \times \tilde{K}(X^W)$ to $\tilde{K}(X^{V \oplus W})$.

A very similar discussion can be carried out for projective spaces. Thus if V is a vector bundle over X let $P = P(V \oplus 1)$ and let H be the standard line-bundle over P. By definition we have a monomorphism

$$H^* \longrightarrow \pi^*(V \oplus 1)$$

when $\pi : P \to X$ is the projection. Hence tensoring with H we get a section of $H \otimes \pi^*(V \oplus 1)$. Projecting onto the first factor gives therefore a natural section

$$s \in \Gamma(H \otimes \pi^* V) \quad .$$

Consider the exterior algebra $\Lambda^*(H \otimes \pi^* V)$. Each component is a vector bundle over P and exterior multiplication by s gives us a complex of vector bundles acyclic outside the subspace where $s = 0$. But this is just the image of the natural cross-section $X \to P$. If we restrict to the complement of $P(V)$ in $P(V \oplus 1)$ then H becomes isomorphic to 1 and we recover the element which defines λ_V (identifying $P(V \oplus 1) - P(V)$ with V in the usual way). This shows that the image of λ_V under the homomorphism

$$\tilde{K}(X^V) = K(P(V \oplus 1), P(V)) \longrightarrow K(P(V \oplus 1))$$

is the alternating sum

$$\Sigma(-1)^i [H]^i [\lambda^i V] \quad .$$

We conclude this section by remarking that everything we have been saying works equally well for G-spaces, G being a finite group. We have only used the basic facts about extensions of homomorphisms etc. which hold equally well for G-bundles. Thus elements of $K_G(X, Y)$ may be represented by G-complexes of vector bundles over X acyclic over Y. In particular the exterior algebra of a G-vector bundle V defines an element

$$\lambda_V \in \tilde{K}_G(X^V)$$

as above.

§2.7. The Thom isomorphism. If $E = \Sigma L_i$ is a decomposable vector bundle over X (i.e., a sum of line-bundles) then we have (2.5.3) determined the structure of K(P(E)) as a K(X)-algebra. Now for any space X we have a canonical isomorphism

$$K^*(X) \cong K(X \times S^1) \quad .$$

Also, if $\pi : X \times S^1 \to X$ is the projection, we have

$$P(E) \times S^1 = P(\pi^* E)$$

and so

$$K^*(P(E)) \cong K(P(\pi^* E)) \quad .$$

Thus replacing X by $X \times S^1$ in (2.5.3) gives at once

PROPOSITION 2.7.1. Let $E = \Sigma L_i$ be a decomposable vector bundle over X. Then $K^*(P(E))$, as a $K^*(X)$-algebra, is generated by [H] subject to the single relation

$$\Pi([L_i][H] - 1) = 0 \quad .$$

Remark: As with (2.5.3) this extends at once to G-spaces giving $K_G^*(P(E))$ as a $K_G^*(X)$-algebra.

Now the Thom space X^E, the one-point compactification of E, may be identified with $P(E \oplus 1)/P(E)$, and at the end of § 2.6 we saw that the image of λ_E in $K(P(E \oplus 1))$

is

$$\Sigma (-1)^i [H]^i [\lambda^i E] = \Pi(1 - [L_i][H]) .$$

Since this element generates (as an ideal) the kernel of

$$K^*(P(E \oplus 1)) \longrightarrow K^*(P(E))$$

we deduce

PROPOSITION 2.7.2. <u>Let</u> E <u>be a decomposable vector bundle over</u> X. <u>Then</u> $\tilde{K}^*(X^E)$ <u>is a free</u> $K^*(X)$<u>-module on</u> λ_E <u>as generator.</u>

<u>Remark:</u> This "Thom isomorphism theorem" for the decomposable case also holds as before for G-spaces. We now show how this fact can be put to use.

COROLLARY 2.7.3. <u>Let</u> X <u>be a G-space such that</u> $K_G^1(X) = 0$ <u>and let</u> E <u>be a decomposable</u> G<u>-vector bundle. Then, if</u> S(E) <u>denotes the sphere bundle, we have an exact sequence</u>

$$0 \longrightarrow K_G^1(S(E)) \longrightarrow K_G^0(X) \xrightarrow{\varphi} K_G^0(X) \longrightarrow K_G^0(S(E)) \longrightarrow 0$$

<u>where</u> φ <u>is multiplication by</u>

$$\lambda_{-1}[E] = \Sigma (-1)^i \lambda^i [E] .$$

Proof: This follows at once by applying (2.7.2) in the exact sequence of the pair $(B(E), S(E))$.

In order to apply this corollary when X = point we need to verify

LEMMA 2.7.4. K_G^1 (point) = 0.

Proof: It is sufficient to show that

$$K_G(S^1) \longrightarrow K_G \text{ (point)}$$

is an isomorphism. But, since G is acting trivially on S^1, we have

$$\begin{aligned} K_G(S^1) &\cong K(S^1) \otimes R(G) \\ &\cong K \text{ (point)} \otimes R(G) \\ &\cong K_G \text{ (point)} . \end{aligned}$$

Thus we can take X = point in (2.7.3). Moreover if we take G abelian then E is necessarily decomposable. Thus we obtain

COROLLARY 2.7.5. Let G be an abelian group, E a G-module. Then we have an exact sequence

$$0 \longrightarrow K_G^1(S(E)) \longrightarrow R(G) \xrightarrow{\varphi} R(G) \longrightarrow K_G^0(S(E)) \longrightarrow 0$$

where φ *is multiplication by*

$$\lambda_{-1}[E] = \Sigma(-1)^i \lambda^i[E] \quad .$$

Suppose in particular that G acts freely on S(E) (it is then necessarily cyclic), so that

$$K^*_G(S(E)) \cong K^*(S(E)/G) \quad .$$

Thus we deduce

COROLLARY 2.7.6. *Let* G *be a cyclic group,* E *a* G-*module with* S(E) G-*free. Then we have an exact sequence*

$$0 \longrightarrow K^1(S(E)/G) \longrightarrow R(G) \xrightarrow{\varphi} R(G) \longrightarrow K^0(S(E)/G) \longrightarrow 0$$

where φ *is multiplication by* $\lambda_{-1}[E]$.

Remark: A similar result will hold for other groups acting freely on spheres once the Thom isomorphism for K_G has been extended to bundles which are not decomposable. However, this will not be done in these notes.

As a special case of (2.7.6) take $G = Z_2$, $E = C^n$ with the (-1) action. Then

$$S(E)/G = P_{2n-1}(R)$$

is real projective space of odd dimension .

$$R(Z_2) = Z[\rho]/\rho^2 - 1$$

$$\lambda_{-1}[E] = (1-\rho)^n \quad .$$

Putting $\sigma = \rho - 1$ so that $\sigma^2 = -2\sigma$ and $\lambda_{-1}[E] = (-\sigma)^n$ we see that $\tilde{K}^0(P_{2n-1}(R))$ is cyclic of order 2^{n-1} while $K^1(P_{2n-1}(R))$ is infinite cyclic. If we compare the sequences for n and $n+1$ we get a commutative diagram

$$\begin{array}{ccccccc} 0 & \longrightarrow & K^1(P_{2n+1}) & \longrightarrow & R(Z_2) & \xrightarrow{(-\sigma)^{n+1}} & R(Z_2) \\ & & \downarrow & & \downarrow {\scriptstyle -\sigma} & & \downarrow {\scriptstyle 1} \\ 0 & \longrightarrow & K^1(P_{2n-1}) & \longrightarrow & R(Z_2) & \xrightarrow{(-\sigma)^{n}} & R(Z_2) \end{array}$$

But in $R(Z_2)$ the kernel of $(-\sigma)^n$ (for $n \geq 1$) is $(2-\sigma)$ and so coincides with the kernel of $-\sigma$. Hence the map

$$K^1(P_{2n+1}) \longrightarrow K^1(P_{2n-1})$$

is zero. From the exact sequences of the pairs (P_{2n+1}, P_{2n}), (P_{2n}, P_{2n-1}) we deduce that

$$K^1(P_{2n+1}) \longrightarrow K^1(P_{2n})$$

is surjective, while

$$K^1(P_{2n}) \longrightarrow K^1(P_{2n-1})$$

is injective. Hence

$$K^1(P_{2n}) = 0 .$$

The exact sequence of the pair (P_{2n+1}, P_{2n}) then shows that

$$K^0(P_{2n+1}) \longrightarrow K^0(P_{2n})$$

is an isomorphism. Summarizing we have established

PROPOSITION 2. 7. 7. _The structure of_ $K^*(P_n(R))$ _is as follows_

$$K^1(P_{2n+1}) = Z$$

$$K^1(P_{2n}) = 0$$

$$\tilde{K}^0(P_{2n+1}) = \tilde{K}^0(P_{2n}) = Z_{2^n} .$$

We leave it as an exercise to the reader to apply (2. 7. 6) to other spaces.

We propose now to proceed to the general Thom isomorphism theorem. It should be emphasized at this point that the methods to be used do _not_ extend to G-bundles. Entirely different methods are needed for G-bundles and we do not discuss them here.

We start with the following general result

THEOREM 2.7.8. *Let $\pi : B \to X$ be a map of compact spaces, and let $\mu_1, \cdots, \mu_n$ be homogeneous elements of $K^*(B)$. Let M^* be the free (Z_2) graded group generated by $\mu_1, \cdots, \mu_n$. Suppose that every point $x \in X$ has a neighborhood U such that for all $V \subset U$, the natural map*

$$K^*(V) \otimes M^* \longrightarrow K^*(\pi^{-1}(V))$$

is an isomorphism. Then, for any $Y \subset X$, the map

$$K^*(X, Y) \otimes M^* \longrightarrow K^*(B, \pi^{-1}(Y))$$

is an isomorphism.

Proof: If $U \subset X$ has the property that, for all $V \subset U$,

$$K^*(V) \otimes M^* \cong K^*(\pi^{-1}(V)) \qquad (1)$$

we shall say that U is good. If U is good then, using exact sequences and the fact that $\otimes M^*$ preserves exactness (M^* being torsion free) we deduce

$$K^*(U, V) \otimes M^* \cong K^*(\pi^{-1}(U), \pi^{-1}(V)) \qquad (2)$$

Here we use of course the compatibility of σ with products (Lemma 2.6.0). What we have to show therefore is

$$X \text{ locally good} \Rightarrow X \text{ good.}$$

Since X is compact it will be enough to show that

$$U_1, U_2 \text{ good} \Rightarrow U_1 \cup U_2 \text{ good.}$$

Now any $V \subset U_1 \cup U_2$ is of the form $V = V_1 \cup V_2$ with $V_i \subset U_i$ (and so V_i is also good). Since

$$V/V_2 = V_1/V_1 \cap V_2$$

it follows that (2) holds for the pair (V, V_2). Since (1) holds for V_2 the exact sequence of (V, V_2) shows that (1) holds for V. Thus $U_1 \cup U_2$ is good and the proof is complete.

COROLLARY 2.7.9. Let $\pi: E \to X$ be a vector bundle, and let H be the usual line bundle over $P(E)$. Then $K^*(P(E))$ is a free $K^*(X)$-module on the generators $1, [H], [H]^2, \cdots, [H]^{n-1}$. $[H]$ satisfies the equation $\Sigma(-1)^i [H]^i [\lambda^i E] = 0$.

Proof: Since E is locally trivial it is in particular locally decomposable.* Hence, by (2.7.1), each point $x \in X$ has a neighborhood U so that for all $V \subset U$, $K^*(P(E|V))$ is a free $K^*(V)$-module on generators $1, [H], \cdots, [H]^{n-1}$. Now apply (2.7.8). The equation for [H] has already been established at the end of §2.6.

COROLLARY 2.7.10. *If* $\pi : E \to X$ *is a vector bundle, and if* F(E) *is the flag bundle of* E *with projection map* $p: F(E) \to X$, *then* $p^* : K^*(X) \to K^*(F(E))$ *is injective.*

Proof: F(E) is the flag bundle over P(E) of a bundle of dimension one less than dim (E). We proceed inductively on dim(E) using (2.7.9).

COROLLARY 2.7.11. (*The Splitting Principle*). *If* $E_1, \cdots, E_n$ *are vector bundles on* X, *then there exist a space* F *and a map* $\pi: F \to X$ *such that*

1) $\pi^* : K^*(X) \to K^*(F)$ *is injective*

2) *Each* $\pi^*(E_i)$ *is a sum of line bundles.*

Proof: We take F to be the flag bundle of $\oplus E_i$.

The importance of the Splitting Principle is clear. It enables

---

* Remark: This is the argument which does not generalize to G-spaces.

us to reduce many problems to the decomposable case.

COROLLARY 2.7.12. (The Thom Isomorphism Theorem). If $\pi : E \to X$ is a vector bundle

$$\Phi : K^*(X) \longrightarrow \tilde{K}^*(X^E)$$

defined by $\Phi(x) = \lambda_E x$ is an isomorphism.

Proof: This follows from (2.7.9) in the same way as (2.7.2) followed from (2.7.1).

We leave the following propositions as exercises for the reader

PROPOSITION 2.7.13. If $\pi : E \to X$ is a vector bundle, $L_1, \cdots, L_n$ the usual line bundles over $F(E)$, then the map defined by $t_i \to [L_i]$ defines an isomorphism of $K^*(X)$ modules

$$K^*(X)[t_1, \cdots, t_n]/I \longrightarrow K^*(F(E))$$

where I is the ideal generated by elements

$$\sigma^1(t_1, \cdots, t_n) - E, \; \sigma^2(t_1, \cdots, t_n) - \lambda^2(E), \cdots, \sigma^n(t_1, \cdots, t_n) - \lambda^n(E)$$

σ^i being the i-th elementary symmetric function.

PROPOSITION 2.7.14. Let $\pi : E \to X$ be an n-dimensional vector bundle and let $G_k(E)$ be the Grassmann bundle (of k-dimensional subspaces) of E. Let F be the induced k-dimensional bundle over $G_k(E)$, F' the quotient bundle $p^*(E)/F$. Then the map defined by $t_i \to \lambda^i(F)$, $s_i \to \lambda^i(F')$ defines an isomorphism of $K^*(X)$-modules

$$K^*(X)[t_1, \cdots, t_k, s_1, \cdots, s_{n-k}]/I \to K^*(G_k(E)) \; ,$$

where I is the ideal generated by the elements

$$\left(\sum_{i+j=\ell} t_i s_j\right) - \lambda^\ell(E) \qquad \text{for all } \ell \, .$$

(Hint: Compare $G_k(E)$ with the flag bundle of E).

In particular, we see that if $G_{n,k}$ is the Grassmann manifold of k-dimensional subspaces of an n-dimensional vector space, $K^*(G_{n,k})$ is torsion free. This also follows from its cell decomposition. By induction we deduce K^* is torsion free for a product of Grassmannians.

THEOREM 2.7.15. Let X be a space such that $K^*(X)$ is torsion free, and let Y be a (finite) cell complex, $Y' \subset Y$ a subcomplex. Then the map

$$K^*(X) \otimes K^*(Y, Y') \longrightarrow K^*(X \times Y, X \times Y')$$

is an isomorphism.

Proof: The theorem holds for Y a ball, Y' its boundary as a consequence of 2.7.2. It thus holds for any (Y, Y') by induction on the number of cells in Y.

COROLLARY 2.7.15. (*The Künneth Theorem*). *Let X be a space such that $K^*(X)$ is a finitely generated abelian group, and let Y be a cell complex. Then there is a natural exact sequence*

$$0 \longrightarrow \sum_{i+j=k} K^i(X) \otimes K^j(X) \longrightarrow K^k(X \times Y)$$

$$\longrightarrow \sum_{i+j=k+1} \operatorname{Tor}(K^i(X), K^j(Y)) \longrightarrow 0$$

where all suffixes are in Z_2.

Proof: Suppose we can find a space Z and a map $f : X \to Z$ such that $K^*(Z)$ is torsion free, and $f^* : K^*(Z) \to K^*(X)$ is surjective. Then from the exact sequence $K^*(Z/X)$ is torsion free. From the last theorem, $K^*(Z \times Y) = K^*(Z) \otimes K^*(Y)$, $K^*((Z/X) \times Y) = K^*(Z/X) \otimes K^*(Y)$. The result will then follow from the exact sequence for the pair $(Z \times Y, X \times Y)$.

We now construct such a map $g : SX \to Z$. Let $a_1, \cdots, a_n$ generate $K^0(X)$, and let $b_1, \cdots, b_m$ generate $K^{-1}(X) = K(SX)$. Then each a_i determines a map $\alpha_i : X \to G_{r_i, s_i}$ for r_i, s_i

suitable, and each b_i a map $\beta_i : SX \to G_{u_i, v_i}$. Let $\alpha : X \to G_{r_1, s_1} \times \cdots \times G_{r_n, s_n} = G'$ be $\alpha_1 \times \cdots \times \alpha_n$, and $\beta : SX \to G_{u_1, v_1} \times \cdots \times G_{u_m, v_m} = G''$ be $\beta_1 \times \cdots \times \beta_m$.

Then

$$\alpha^* : K^0(G') \longrightarrow K^0(X) \quad \text{is surjective}$$

$$\beta^* : K^0(G'') \longrightarrow K^0(SX) \quad \text{is surjective.}$$

Thus, if $f : (S\alpha) \times \beta : SX \to (SG') \times G'' = G$

$$f^* : K^*(G) \longrightarrow K^*(SX) \quad \text{is surjective,}$$

and $K^*(G)$ is torsion free as required. This proves the formula for SX and this is equivalent to the formula for X.

We next compute the rings $K^*(U(n))$, where $U(n)$ is the unitary group on n variables. Now for any compact Lie group G we can consider representations $\rho : G \to GL(m, C)$ as defining elements $[\rho] \in K^1(G)$: we simply regard ρ as a map and disregard its multiplicative properties. Suppose now that α, β are two representations $G \to GL(m, C)$ which agree on the closed subgroup H. Then we can define a map

$$\gamma : G/H \to GL(m, C)$$

by $\gamma(gH) = \alpha(g)\beta(g)^{-1}$. This is well-defined because of the multiplicative properties of α, β. The map γ defines an element

$[\gamma] \in K^1(G/H)$ whose image in $K^1(G)$ is just $[\alpha] - [\beta]$. As a particular case of this we take

$$G = U(n), \quad H = U(n-1), \quad G/H = S^{2n-1} \quad .$$

For α, β we take the representations of G on the even and odd parts of the exterior algebra $\Lambda^*(C^n)$, and we identify these two parts by exterior multiplication with the n-th basic vector e_n of C^n. Since $U(n-1)$ keeps e_n fixed this identification is compatible with the action of $U(n-1)$. We are thus in the situation being considered and so we obtain an element

$$[\gamma] \in K^1(S^{2n-1}) \quad .$$

If we pass to the isomorphic group $\tilde{K}(S^{2n})$ we see from its definition that $[\gamma]$ is just the basic element

$$\lambda_{C^n} \in \tilde{K}(S^{2n})$$

constructed earlier from the exterior algebra. Thus $[\gamma]$ is a generator of $K^1(S^{2n-1})$, and its image in $K^1(U(n))$ is $\Sigma(-1)^i[\lambda^i]$, where the λ^i are the exterior power representations. With this preliminary discussion we are now ready to prove:

THEOREM 2.7.17. $K^*(U(n))$ *is the exterior algebra generated by* $[\lambda^1], \cdots, [\lambda^n]$, *where* λ^i *is the i-th exterior power representation of* $U(n)$.

Proof: We proceed by induction on n. Consider the mapping

$$U(n) \longrightarrow U(n)/U(n-1) = S^{2n-1} \quad .$$

Since the restriction of λ^i to $U(n-1)$ is $\mu^i \oplus \mu^{i-1}$, where μ^i denotes the i-th exterior power representation of $U(n-1)$, the inductive hypothesis together with (2.7.8) imply that $K^*(U(n))$ is a free $K^*(S^{2n-1})$-module generated by the monomials in $[\lambda^1], \cdots, [\lambda^{n-1}]$. But $K^*(S^{2n-1})$ is an exterior algebra on one generator $[\gamma]$ whose image in $K^*(U(n))$ is

$$\sum_{i=0}^{n} (-1)^i [\lambda^i] \quad ,$$

as shown above. Hence $K^*(U(n))$ is the exterior algebra on $[\lambda^1], \cdots, [\lambda^n]$ as required.

CHAPTER III. Operations.

§1. Exterior Powers. By an operation F in K-theory, we shall mean a natural transformation $F_X : K(X) \to K(X)$. That is, for every space X, there is a (set) map $F_X : K(X) \to K(X)$, and if $f : X \to Y$ is any continuous map, $F_X f^* = f^* F_Y$.

Suppose that F and G are two operations which have the property that $F([E] - n) = G([E] - n)$ whenever E is a sum of line bundles and n is an integer. Then $F(x) = G(x)$ for all $x \in K(X)$, as we see immediately from the splitting principle of the last chapter.

There are various ways in which one can define operations using exterior power operations. The first of these which we shall discuss is due to Grothendieck.

If V is a vector bundle over a space X, we define $\lambda_t[V] \in K(X)[[t]]$ to be the power series

$$\sum_{i=0} t^i[\lambda^i(V)] \quad .$$

The isomorphism

$$\lambda^k(V \oplus W) \cong \sum_{i+j=k} \lambda^i(V) \otimes \lambda^j(W)$$

gives us the formula

$$\lambda_t[V \oplus W] = \lambda_t[V]\lambda_t[W]$$

for any two bundles V, W. For any W the power series $\lambda_t[W]$ is a unit in $K(X)[[t]]$, because it has constant leading term 1.

Thus we have a homomorphism

$$\lambda_t : \mathrm{Vect}(X) \longrightarrow 1 + K(X)[[t]]^+$$

of the additive semi-group Vect(X) into the multiplicative group of power series over K(X) with constant term 1. By the universal property of K(X) this extends uniquely to a homomorphism

$$\lambda_t : K(X) \to 1 + K(X)[[t]]^+ \quad .$$

Thus, taking the coefficient of t^i we have operations

$$\lambda^i : K(X) \longrightarrow K(X) \quad .$$

Explicitly therefore

$$\lambda_t([V] - [W]) = \lambda_t[V]\lambda_t[W]^{-1} \quad .$$

In a very similar way we can treat the symmetric powers $S^i(V)$. Since

$$S^k(V \oplus W) \cong \sum_{i+j=k} S^i(V) \otimes S^j(W)$$

we obtain a homomorphism

$$S_t : K(X) \longrightarrow 1 + K(X)[[t]]^+$$

whose coefficients define the operations

$$S^i : K(X) \longrightarrow K(X) \quad .$$

Notice that if L is a line bundle,

$$\begin{aligned} \lambda_t(L) &= 1 + tL \\ S_t(L) &= 1 + tL + t^2L + \cdots \\ &= (1 - tL)^{-1} \quad . \end{aligned}$$

Thus

$$\lambda_{-t}(L)S_t(L) = 1 \quad .$$

Thus, if V is a sum of line bundles, $\lambda_{-t}[V]S_t[V] = 1$. Therefore, for any $x \in K(X)$, $\lambda_{-t}(x)S_t(x) = 1$, and so

$$\lambda_t([V] - [W]) = \lambda_t[V]S_{-t}[W]$$

that is,

$$\lambda^k([V] - [W]) = \sum_{i+j=k} (-1)^j \lambda^i[V]S^j[W] \quad .$$

This gives us an explicit formula for the operations λ^i in terms of operations on bundles.

Now recall that, for any bundle E, $\dim E_x$ is a locally constant function of x. Since X is assumed compact

$$\dim E = \sup_{x \in X} \dim E_x$$

is finite. The exterior powers have the basic property that

$$\lambda^i E = 0 \qquad \text{if} \qquad i > \dim E .$$

Let us call an element of $K(X)$ <u>positive</u> (written $x \geq 0$) if it is represented by a genuine bundle, i. e., if it is in the image of $\mathrm{Vect}(X)$. Then

$$x \geq 0 \implies \lambda_t(x) \in K(X)[t] .$$

For many problems it is not $\dim E$ which is important but another integer defined as follows. First let us denote by rank E the bundle whose fibre at x is $C^{d(x)}$ where $d(x) = \dim E_x$: if X is connected then rank E is just the trivial bundle of dimension equal to $\dim E$. Then $E \to \operatorname{rank} E$ induces an (idempotent) ring endomorphism

$$\operatorname{rank}: K(X) \longrightarrow K(X)$$

which is frequently referred to as the <u>augmentation</u>. The kernel of this endomorphism is an ideal denoted by $K_1(X)$. For a connected space with base-point we clearly have

$$K_1(X) = \widetilde{K}(X) \quad .$$

For any $x \in K(X)$ we have

$$x - \text{rank}\, x \in K_1(X) \quad .$$

Now define $\dim_K x$, for any $x \in K(X)$, to be the least integer n for which

$$x - \text{rank}\ x + n \geq 0$$

since every element of $K(X)$ can be represented in the form $[V] - n$ for some bundle V it follows that $\dim_K x$ is finite for all $x \in K(X)$. For a vector bundle E we clearly have

$$\dim_K[E] \leq \dim E \quad .$$

Notice that

$$\dim_K \cdot x = \dim_K x_1$$

where $x_1 = x - \text{rank}\, x$, so that $\dim_K$ is essentially a function on the quotient $K_1(X)$ of $K(X)$.

It is now convenient to introduce operations γ^i which have the same relation to $\dim_K$ as the λ^i have to the dimension of bundles. Again following Grothendieck we define

$$\gamma_t(x) = \lambda_{t/1-t}(x) \in K(X)[[t]]$$

so that $\gamma_t(x+y) = \gamma_t(x)\gamma_t(y)$. Thus for each i we have an operation

$$\gamma^i : K(X) \to K(X) \quad .$$

The γ^i are linear combinations of the λ^j for $j \leq i$ and vice-versa, in view of the formula

$$\lambda_s(x) = \gamma_{s/1+s}(x)$$

obtained by putting $s = t/1 - t$, $t = s/1 + s$. Note that

$$\gamma_t(1) = (1-t)^{-1}$$

and for a line-bundle L

$$\gamma_t([L]-1) = 1 + t([L]-1) \quad .$$

PROPOSITION 3.1.1. <u>Let</u> $x \in K_1(X)$, <u>then</u> $\gamma_t(x)$ <u>is a polynomial of degree</u> $\leq \dim_K x$.

<u>Proof:</u> Let $n = \dim_K x$, so that $x + n \geq 0$. Thus $x + n = [E]$ for some vector bundle E. Moreover $\dim E = n$ and so

$$\lambda^i(E) = 0 \qquad \text{for} \quad i > n \quad .$$

Thus $\lambda_t(x+n)$ is a polynomial of degree $\leq n$. Now

$$\gamma_t(x) = \gamma_t(x+n)\gamma_t(1)^{-n}$$
$$= \gamma_{t/1-t}(x+n)(1-t)^n$$
$$= \sum_{1=0}^{n} \lambda^i(x+n)t^i(1-t)^{n-i}$$

and so is a polynomial of degree $\leq n$ as stated.

We now define $\dim_\gamma x$ to be the largest integer n such that $\gamma^n(x - \text{rank } x) \neq 0$, and we put

$$\dim_K X = \sup_{x \in K(X)} \dim_K x$$

$$\dim_\gamma X = \sup_{x \in K(X)} \dim_\gamma x \quad .$$

By (3.1.1) we have

$$\dim_\gamma x \leq \dim_K x \, , \quad \dim_\gamma X \leq \dim_K X \quad .$$

We shall show that, under mild restrictions, $\dim_K X$ is finite. For this we shall need some preliminary lemmas on symmetric functions.

LEMMA 3.1.2. Let $x_1, \cdots, x_n$ be indeterminates. Then any homogeneous polynomial in $Z[x_1, \cdots, x_n]$ of degree $> n(n-1)$ lies in the ideal generated by the symmetric functions of $(x_1, \cdots, x_n)$ of positive degree.

Proof: Let $\sigma_i(x_1, \cdots, x_n)$ be the i-th elementary symmetric function. Then the equation

$$x^n - \sigma_1 x^{n-1} + \sigma_2 x^{n-2} + \cdots + (-1)^n \sigma_n = 0$$

has roots $x = x_i$. Thus x_i^n is in the ideal generated by $\sigma_1, \cdots, \sigma_n$. But any monomial in $x_1, \cdots, x_n$ of degree $> n(n-1)$ is divisible by x_i^n for some i and so is also in this ideal.

LEMMA 3.1.3. Let $x_1, \cdots, x_n$, $y_1, \cdots, y_m$ be indeterminates and let

$$a_i = \sigma_i(x_1, \cdots, x_n) \qquad b_i = \sigma_i(y_1, \cdots, y_m)$$

be the elementary symmetric functions. Let I be any ideal in $Z[a,b]$, J its extension in $Z[x,y]$. Then

$$J \cap Z[a,b] = I .$$

Proof: It is well-known that $Z[x]$ is a free $Z[a]$-module with basis the monomials

$$x^{\underline{r}} = x_1^{r_1} x_2^{r_2} \cdots x_{n-1}^{r_{n-1}} \qquad 0 \le r_i \le n - i .$$

Hence $Z[x,y] = Z[x] \otimes Z[y]$ is a free module over $Z[a,b] = Z[a] \otimes Z[b]$ with basis the monomials $x^{\underline{r}} y^{\underline{s}}$. Then the ideal $J \subset Z[x,y]$ consists of all elements f of the form

$$f = \sum f_{\underline{r},\underline{s}} x^{\underline{r}} y^{\underline{s}} \qquad \text{with} \quad f_{\underline{r},\underline{s}} \in I .$$

Since the $x^{\underline{r}} y^{\underline{s}}$ are a free basis f belongs to $Z[a,b]$ if and only if $f_{\underline{r},\underline{s}} = 0$ for $\underline{r},\underline{s} \neq (0,0)$ in which case

$$f = f_{0,0} \in I .$$

Thus $J \cap Z[a,b] = I$ as stated.

Remark: This lemma is essentially an algebraic form of the splitting principle since it asserts that we can embed $Z[a,b]/I$ in $Z[x,y]/J$. It is of course purely formal in character and it seems preferable to use this rather than the topological splitting principle whenever we are dealing with formal algebraic results. The topological splitting principle depends of course on the periodicity theorem and should only be used when we are dealing with properties that lie at that depth.

LEMMA 3.1.4. Let K be a commutative ring (with 1) and suppose

$$a(t) = 1 + a_1 t + a_2 t^2 + \cdots + a_n t^n$$

$$b(t) = 1 + b_1 t + b_2 t^2 + \cdots + b_m t^m$$

are elements of K[t] *such that*

$$a(t)b(t) = 1 .$$

Then there exists an integer $N = N(n, m)$ *so that any monomial*

$$a_1^{r_1} a_2^{r_2} \cdots a_n^{r_n}$$

of weight $\Sigma\, jr_j > N$ *vanishes*.

Proof: Passing to the universal situation it is sufficient to prove that if $a_1, \cdots, a_n, b_1, \cdots, b_m$ are indeterminates, then any monomial α in the a_i of weight $\geq N$ lies in the ideal I generated by the elements

$$c_k = \sum_{i+j=k} a_i b_j \qquad k = 1, \cdots, mn (a_0 = b_0 = 1) .$$

By (3.1.3), introducing indeterminates $x_1, \cdots, x_n, y_1, \cdots, y_m$, it is sufficient to prove that α belongs to the extended ideal J. But c_k is just the k-th elementary symmetric function of the $(m + n)$ variables $x_1, \cdots, x_n, y_1, \cdots, y_m$. The result now follows by applying (3.1.2) with $N = (m + n)(m + n - 1)$.

Remark: The value for $N(m,n)$ obtained in the above proof is not best possible. It can be shown by more detailed arguments that the best possible value is mn .

We now apply these algebraic results:

PROPOSITION 3.1.5. Let $x \in K_1(x)$. Then there exists an integer N , depending on x , such that any monomial

$$\gamma^{i_1}(x)\,\gamma^{i_2}(x)\ldots\gamma^{i_k}(x)$$

of weight $\Sigma_{j=1}^{k} i_j > N$ is equal to zero.

Proof: We apply (3.1.4) to the polynomials $\gamma_t(x)$, $\gamma_t(-x)$. Note therefore, that N depends on $\dim_\gamma x$, $\dim_\gamma(-x)$.

Since $\gamma^1(x) = x$ we deduce:

COROLLARY 3.1.6. Any $x \in K_1(X)$ is nilpotent.

If we define the degree of each γ^i to be one, then for any monomial in the γ^i we have

$$\text{weight} \geq \text{degree} .$$

In view of (3.1.5), therefore, all monomials in $\gamma^i(x)$ of sufficiently high degree are zero if $x \in K_1(X)$. Thus we can apply a formal power series* in the γ^i to any $x \in K_1(X)$. Let us denote by

---

* As usual a formal power series means a sum $f = \Sigma f_n$ where f_n is a homogeneous polynomial of degree n (and so involves only a finite number of the variables).

$Op(K_1, K)$ the set of all operations $K_1 \to K$. This has a ring structure induced by the ring structure of K (addition and multiplication of values). Then by what we have said we obtain a ring homomorphism

$$\varphi : Z[[\gamma^1, \cdots, \gamma^n, \cdots]] \longrightarrow Op(K_1, K) .$$

THEOREM 3.1.7.

$$\varphi : Z[[\gamma^1, \cdots, \gamma^n, \cdots]] \longrightarrow Op(K_1, K)$$

is an isomorphism.

Proof: Let $Y_{n,m}$ be the product of n copies of $P_m(C)$. Using the base point $P_0(C)$ of $P_m(C)$ the $Y_{n,m}$ form a direct system of spaces with inclusions

$$Y_{n,m} \longrightarrow Y_{n',m'} \qquad \text{for } n' \geq n, \ m' \geq m .$$

Then $K(Y_{n,m})$ is an inverse system of groups with

$$K(Y_{n,m}) = Z[x_1, \cdots, x_n]/(x_1^{m+1}, \cdots, x_n^{m+1})$$

$$\varprojlim_m K(Y_{n,m}) = Z[[x_1, \cdots, x_n]]$$

$$\varprojlim_{m,n} K(Y_{n,m}) = \varprojlim_n Z[[x_1, \cdots, x_n]] .$$

Any operation will induce an operation on the inverse limits. Hence we can define a map

$$\psi : Op(K_1, K) \longrightarrow \varprojlim_n Z[[x_1, \cdots, x_n]]$$

by $\psi(f) = \varprojlim f(x_1 + x_2 + \cdots + x_n)$. Since, in $K(Y_{n,m})$ we have

$$\gamma_t(x_1 + x_2 + \cdots + x_n) = \prod_{i=1}^{n} (1 + x_i t)$$

it follows that

$$\psi\varphi(\gamma^i) = \varprojlim_n \sigma_i(x_1, \cdots, x_n)$$

where σ_i denotes the i-th elementary symmetric function. In particular, therefore $\psi\varphi$ is injective and so φ is injective. Moreover the image of $\psi\varphi$ is

$$Z[[\sigma_1, \cdots, \sigma_n]]$$

which is the same as

$$\varprojlim_n Z[[x_1, \cdots, x_n]]^{S_n}$$

where $[\ ]^{S_n}$ denotes the subring of invariants under the symmetric group S_n . But, for all $f \in Op(K_1, K)$,

$$\psi(f) = \varprojlim f(x_1 + \cdots + x_n)$$

lies in this group. In other words

$$\operatorname{Im} \psi\varphi = \operatorname{Im} \psi \ .$$

To complete the proof it remains now to show that ψ is injective. Suppose then that $\psi(f) = 0$. Since any line bundle over a space X is induced by a map into some $P_n(C)$ it follows that

$$f([E] - n) = 0$$

whenever E is a sum of n line-bundles. By the splitting principle this implies that

$$f(x) = 0 \qquad\qquad \text{for all} \quad x \in K_1 \ ,$$

i.e., f is the zero operation, as required.

Let us define $H^0(X, Z)$ to be the ring of all continuous maps $X \to Z$. Then we have a direct sum decomposition of groups

$$K(X) = K_1(X) \oplus H^0(X, Z)$$

determined by the rank homomorphism. It is easy to see that there are no non-zero natural homomorphisms

$$H^0(X, Z) \longrightarrow K_1(X)$$

and so $Op(K) = Op(K, K)$ differs from $Op(K_1, K)$ only by $Op(H^0(Z))$ which is the ring of all maps $Z \to Z$. Thus (3.1.7) gives essentially a complete description of $Op(K)$.

We turn now to a discussion of finiteness conditions on $K(X)$. First we deal with $H^0(X, Z)$.

PROPOSITION 3.1.8. *The following are equivalent*

(A) $H^0(X, Z)$ *is a Noetherian ring*

(B) $H^0(X, Z)$ *is a finite Z-module*.

Proof: (B) implies (A) trivially. Suppose therefore that $H^0(X, Z)$ is Noetherian. Assume if possible that we can find a strictly decreasing infinite chain of components (open and closed sets) of X

$$X = X_0 \supset X_1 \supset \cdots \supset X_n \supset X_{n+1} \supset \cdots \quad .$$

Then for each n we can find a continuous map $f_n : X \to Z$ so that

$$f_n(X_{n+1}) = 0$$
$$f_n(X_n - X_{n+1}) = 1 \quad .$$

Consider the ideal I of $H^0(X, Z)$ consisting of maps $f : X \to Z$ such that $f(X_n) = 0$ for some n. Since $H^0(X, Z)$ is Noetherian I is finitely generated and hence there exists N so that

$$f(X_N) = 0 \qquad \text{for all } f \in I .$$

But this is a contradiction because

$$f_N \in I , \quad f_N(X_N) \neq 0 .$$

Thus X has only a finite number of components, so that

$$X = \sum_{i=1}^{n} X_i$$

with X_i connected. Hence $H^0(X, Z)$ is isomorphic to Z^n.

Passing now to $K(X)$ we have

PROPOSITION 3.1.9. The following are equivalent

(A) $K(X)$ is a Noetherian ring

(B) $K(X)$ is a finite Z-module .

Proof: Again assume (A), then $H^0(X, Z)$ which is a quotient ring of $K(X)$ is also Noetherian. Hence by (3.1.8), $H^0(X, Z)$ is a finite Z-module. Now $K_1(X)$ is an ideal of $K(X)$ consisting of nilpotent elements (3.1.6). Since $K(X)$ is Noetherian it follows that $K_1(X)$ is a nilpotent ideal. For brevity put $I = K_1(X)$. Then $I^n = 0$ for some n and the I^m/I^{m+1}, $m = 0, 1, \ldots, n-1$ are all finite modules over $K/I = H^0(X, Z)$. Hence $K(X)$ is a finite

$H^0(X, Z)$-module and so also a finite Z-module.

Examples of spaces X for which $K(X)$ is a finite Z-module are cell-complexes.

Let us now define a filtration of $K(X)$ by the subgroups $K_n^\gamma(X)$ generated by all monomials

$$\gamma^{i_1}(x_1)\,\gamma^{i_2}(x_2)\cdots\gamma^{i_k}(x_k)$$

with $\Sigma_{j=1}^k i_j \geq n$ and $x_i \in K_1(X)$. Since $\gamma^1(x) = x$, we have $K_1^\gamma = K_1$. If $x \in K_n^\gamma(X)$ we say that x has γ-filtration $\geq n$ and write $F_\gamma(x) \geq n$.

PROPOSITION 3.1.10. <u>Assume</u> $K(X)$ <u>is a finite</u> Z-<u>module. Then for some</u> n

$$K_n^\gamma(X) = 0 .$$

<u>Proof</u>: Let $x_1, \cdots, x_s$ be generators of $K_1(X)$ and let $N_j = N(x_j)$ be the integers given by (3.1.5). Because of the formula

$$\gamma_t(a + b) = \gamma_t(a)\,\gamma_t(b)$$

it will be sufficient to show that there exists N so that all monomials in the $\gamma^i(x_j)$ of total weight $> N$ are zero. But taking $N = \Sigma_{j=1}^s N_j$ we see that any such monomial must, for some j, have weight $> N_j$

in the $\gamma^i(x_j)$. Hence by (3.1.5) this monomial is zero.

COROLLARY 3.1.11. Assume $K(X)$ is a finite Z-module. Then $\dim_\gamma X$ is finite.

We call the reader's attention to certain further properties of the operations γ^i.

PROPOSITION 3.1.12. If V is a bundle of dimension n, $\lambda_{-1}[V] = (-1)^n \gamma^n([V] - n)$. Thus $\widetilde{K}^*(X^V)$ is a free $K^*(X)$ module generated by $\gamma^n([V] - n)$.

PROPOSITION 3.1.13. There exist polynomials P_i, Q_{ij} such that for all x, y

$$\gamma^i(xy) = P_i(\gamma^1(x), \gamma^1(y), \gamma^2(x), \gamma^2(y), \ldots, \gamma^i(x), \gamma^i(y))$$

$$\gamma^i(\gamma^j(x)) = Q_{ij}(\gamma^1(x), \ldots, \gamma^{i+j}(x)) .$$

We leave these proofs to the reader, who may verify them easily by use of the splitting principle.

§2. The Adams Operations. We shall now separate out for special attention some operations with particularly pleasing properties. These were introduced by J. F. Adams. We define $\psi^0(x) = \text{rank}(x)$. In the ring $K(X)[[t]]$ we define $\psi_t(x) = \Sigma_{i=0} t^i \psi^i(x)$ by

$$\psi_t(x) = \psi^0(x) - t \frac{d}{dt}(\log \lambda_{-t}(x)) .$$

Notice that since all of the coefficients of this power series are integers, this definition makes sense.

PROPOSITION 3.2.1. For any $x, y \in K(X)$

1) $\psi^k(x + y) = \psi^k(x) + \psi^k(y)$ for all k

2) If x is a line bundle, $\psi^k(x) = x^k$.

3) Properties 1 and 2 uniquely determine the operations ψ^k.

Proof: $\psi_t(x + y) = \psi_t(x) + \psi_t(y)$, so that $\psi^k(x + y) = \psi^k(x) + \psi^k(y)$ for each k.

If x is a line bundle, $\lambda_{-t}(x) = 1 - tx$, so that

$$\frac{d}{dt}(\log(1 - tx)) = \frac{-x}{1 - tx}$$

$$= -x - tx^2 - t^2x^3 - \cdots .$$

Thus $\psi_t(x) = 1 + tx + t^2x^2 + \cdots$.

The last part follows from the splitting principle.

PROPOSITION 3.2.2. For any $x, y \in K(X)$

1) $\psi^k(xy) = \psi^k(x)\,\psi^k(y)$ for all k

2) $\psi^k(\psi^\ell(x)) = \psi^{k\ell}(x)$ for all k, ℓ .

3) If p is prime, $\psi^p(x) \equiv x^p \bmod p$

4) If $u \in \widetilde{K}(S^{2n})$, $\psi^k(u) = k^n u$ for all k .

Proof: The first two assertions follow immediately from the last proposition and the splitting principle. Also, from the splitting principle, $\psi^p(x) = x^p + pf(\lambda^1(x), \cdots, \lambda^p(x))$, where f is some polynomial with integral coefficients. Finally, if h is the generator of $\widetilde{K}(S^2)$, $\psi^k(h) = kh$. Since $S^{2n} = S^2 \wedge \cdots \wedge S^2$, and $\widetilde{K}(S^{2n})$ is generated by $h \otimes h \otimes \cdots \otimes h$, the last assertion follows from the first.

We next give an application of the Adams operations ψ^k . Suppose that $f : S^{4n-1} \to S^{2n}$ is any map. We define the Hopf invariant $H(f)$ as follows. Let X_f be the mapping cone of f . Let $i : S^{2n} \to X_f$ be the inclusion, and let $j : X_f \to S^{4n}$ collapse S^{2n} . Let u be the generator of $\widetilde{K}(S^{4n})$. From the exact sequence we see that there is an element $x \in \widetilde{K}(X_f)$ such that $i^*(x)$ generates $\widetilde{K}(S^{2n})$. $\widetilde{K}(X_f)$ is the free abelian group generated by

x and $y = j^*(u)$. Since $(i^*(x))^2 = 0$, $x^2 = Hy$ for some H. This integer H we define as the Hopf invariant of f. Clearly, up to a minus sign, $H(f)$ is well defined. The following theorem was first established by J. F. Adams by cohomological methods.

THEOREM 3.2.3. *If* $H(f)$ *is odd, then* $n = 1, 2,$ *or* 4.

Proof: Let $\psi^2(x) = 2^n x + ay$, $\psi^3(x) = 3^n x + by$. Since $\psi^2(x) \equiv x^2 \bmod 2$, a is odd. $\psi^k(y) = j^*(\psi^k(u)) = k^{2n} y$. Thus, we see that

$$\psi^6(x) = \psi^3(\psi^2(x)) = 6^n x + (2^n b + 3^{2n} a)y$$

$$\psi^6(x) = \psi^2(\psi^3(x)) = 6^n x + (2^{2n} b + 3^n a)y.$$

Thus $2^n b + 3^{2n} a = 2^{2n} b + 3^n a$, or $2^n(2^n - 1)b = 3^n(3^n - 1)a$. Since a is odd, 2^n divides $3^n - 1$, which by elementary number theory can happen only if $n = 1, 2,$ or 4.

If $n = 1, 2,$ or 4, the Hopf maps determined by considering S^{4n-1} as a subspace of the non-zero vectors in 2-dimensional complex, quaternionic, or Cayley space, and S^{2n} as the complex, quaternionic, or Cayley projective line all have Hopf invariant one. We leave the verification to the reader.

PROPOSITION 3.2.4. *Let* $x \in K(X)$ *be such that* $F_\gamma(x) \geq n$. *Then for any* k *we have*

$$F_\gamma(\psi^k(x) - k^n x) \geq n + 1 .$$

<u>Proof:</u> If $n = 0$ we have

$$\psi^k(x) = \psi^k(\text{rank } x + x_1) = \text{rank } x + \psi^k x_1 \quad .$$

Here x_1 and so $\psi^k x_1$ are in $K_1(X)$. Thus

$$\psi^k x - x = \psi^k x_1 - x_1 \in K_1(X) = K_1^\gamma(X) \quad .$$

Consider now $n > 0$. Since ψ^k is a ring homomorphism it is sufficient to prove that the composition $\psi^k \circ \gamma^n - k^n \gamma^n$ (where $\psi^k \in Op(K)$, $\gamma^n \in Op(K_1, K)$) is equal to a polynomial in the γ^i in which each term has weight $\geq n+1$. As in (3.1.7) we have isomorphisms

$$Z[[\gamma^1, \cdots]] \cong Op(K_1, K) \cong \varprojlim_m Z[x_1, \cdots, x_m]^{Sm}$$

in which γ^i corresponds to i-th elementary symmetric function σ_i of the x_j. Now

$$\psi^k(x_i) = (1 + x_i)^k - 1$$

and so

$$\psi^k(\sigma_n(x_1, \cdots)) = \sigma_n((1 + x_1)^k - 1, \cdots)$$

$$= k^n \sigma_n(x) + f$$

where f is a polynomial in the σ_i of weight $\geq n+1$. Since

$\psi^k \circ \gamma^n$ corresponds to $\psi^k(\sigma_n)$ by the above isomorphisms the proposition is established.

Iterating (3. 2. 4) we obtain:

COROLLARY 3. 2. 5. *If* $K^{\gamma}_{n+1}(X) = 0$,

$$\left[\prod_{m=0}^{n} \left(\psi^{k_m} - (k_m)^m\right)\right] = 0$$

for any sequence of non-negative integers $k_0, k_1, \cdots, k_n$.

By (3. 1. 10) we can apply 3. 2. 5 in particular whenever $K(X)$ is a finite Z-module.

Notice that ψ^k acts as a linear transformation on the vector space $K(X) \otimes Q$. Taking $k_m = k$ for all m in (3. 2. 5) we see that

$$\prod_{m=0}^{n} (\psi^k - k^m) = 0 \qquad \text{on } K(X) \otimes Q .$$

Thus the eigenvalues of each ψ^k are powers of k not exceeding k^n. Let $V_{k,i}$ denote the eigenspace of ψ^k corresponding to the eigenvalue k^i (we may have $V_{k,i} = 0$). Then if $k > 1$, we have an orthogonal decomposition of the identity operator 1 of $K(X) \otimes Q$:

$$1 = \sum \Pi_i \quad , \quad \Pi_i = \prod_{m \neq i} (\psi^k - k^m)/(k^i - k^m) \quad .$$

Thus $K(X) \otimes Q$ is the direct sum of the $V_{k,i}$. Now put in (3. 2. 5),

$$k_i = \ell , \quad k_m = k \qquad \text{for } m \neq i$$

and we see that

$$(\psi^\ell - \ell^i) V_{k,i} = 0$$

and so $V_{k,i} \subset V_{\ell,i}$. Hence we deduce

PROPOSITION 3. 2. 6. Assume $K(X)$ has finite γ-filtration and let $V_{k,i}$ denote the eigenspace of ψ^k on $K(X) \otimes Q$ corresponding to the eigenvalue k^i . Then if $k, \ell > 1$ we have

$$V_{k,i} = V_{\ell,i} \quad .$$

Since the subspace $V_{k,i}$ does not depend on k (for $k > 1$) we may denote it by a symbol independent of k . We shall denote it by $H^{2i}(X; Q)$ and call it the 2i-th Betti group of X . From (3. 2. 4) it follows that the eigenvalue $k^0 = 1$ occurs only in $H^0(X, Z) \otimes Q$. Thus our notation is consistent in that

$$H^0(X, Z) \otimes Q = H^0(X ; Q) \quad .$$

We define the odd Betti groups by

$$H^{2m+1}(X;Q) = H^{2m+2}(SX^+;Q)$$

where $X^+ = X \cup$ point and S denotes reduced suspension. If the spaces involved are finite-dimensional we put

$$B_k = \dim_Q H^k(X;Q)$$

and the Euler characteristic $E(X)$ is defined by

$$E(X) = \Sigma(-1)^k B_k = \dim_Q(K^0(X) \otimes Q) - \dim_Q(K^1(X) \otimes Q).$$

Note that the Künneth formula (when applicable) implies

$$E(X \times Y) = E(X)E(Y) .$$

The following proposition is merely a reformulation of (3.2.4) in terms of the notation just introduced:

PROPOSITION 3.2.7.

$$K_n^\gamma(X) \otimes Q = \sum_{m \geq n} H^{2m}(X;Q)$$

and so

$$\{K_n^\gamma(X)/K_{n+1}^\gamma(X)\} \otimes Q \cong H^{2n}(X;Q).$$

Since $\psi^k u = ku$ for the generator u of $\tilde{K}(S^2)$ it follows that

$$\psi^k \beta(x) = k\beta\psi^k(x)$$

where $\beta: K(X) \to K^{-2}(X)$ is the periodicity isomorphism. Thus β induces an isomorphism

$$H^{2m}(X ; Q) \cong H^{2m+2}(S^2X^+; Q) .$$

From the way the odd Betti groups were defined it follows that, for all k

$$H^k(X ; Q) \cong H^{k+1}(SX^+; Q) . \tag{3.2.8}$$

If we now take the exact K-sequence of the pair X, A, tensor with Q, decompose under ψ^k and use (3.2.8) we obtain:

PROPOSITION 3.2.9. <u>If $A \subset X$, and if both $K^*(X)$, $K^*(A)$ are finite Z-modules the exact sequence</u>

$$\cdots \longrightarrow K^{i-1}(A) \xrightarrow{\delta} K^i(X,A) \longrightarrow K^i(X) \longrightarrow K^i(A) \xrightarrow{\delta} \cdots$$

induces an exact sequence

$$\cdots \longrightarrow H^{i-1}(A;Q) \xrightarrow{\delta} H^i(X,A;Q) \longrightarrow H^i(X;Q) \longrightarrow H^i(A;Q) \xrightarrow{\delta} \cdots$$

We next give a second application of the operations ψ^k. Since $P_n(C)/P_{n-1}(C)$ is the sphere S^{2n}, we have an inclusion of S^{2n} into $P_{n+k}(C)/P_{n-1}(C)$ for all k. We should like to know for which values of n and k, S^{2n} is a retract of $P_{n+k}(C)/P_{n-1}(C)$. That is, we should like to know when can there exist a map $f : P_{n+k}(C)/P_{n-1}(C) \to S^{2n}$ which is the identity on S^{2n}. We shall obtain certain necessary conditions on n and k for such an f to exist.

THEOREM 3.2.10. Assume a retraction

$$f : P_{n+k}(C)/P_{n-1}(C) \longrightarrow P_n(C)/P_{n-1}(C) = S^{2n}$$

exists. Then the coefficients of x^i for $i \leq k$ in $\left(\frac{\log 1+x}{x}\right)^n$ are all integers.

Proof: Let ξ be the usual line-bundle over P_{n+k} and let $x = \xi - 1$. Then $K(P_{n+k})$ is a free abelian group on generators x^s, $0 \leq s \leq n+k$, and we may identify $K(P_{n+k}, P_{n-1})$ with the subgroup generated by x^s with $n \leq s \leq n+k$. In $K(P_{n+k}) \otimes Q$ put $y = \log(1+x)$, so that $\xi = e^y$. Then

$$e^{ry} = \xi^r = \psi^r(e^y) = e^{\psi^r(y)} ,$$

so that $\psi^r(y) = ry$. Thus $H^{2s}(P_{n+k}/P_{n-1}; Q)$, for $n \leq s \leq n+k$

is a one-dimensional space generated by y^s . Now let u generate $\widetilde{K}(S^{2n})$, and let

$$f^*(u) = \sum_{i=n}^{n+k} a_i x^i \quad .$$

Since f is a retract we have $a_n = 1$. Since $\psi^k u = k^n u$, $f^*(u)$ must be a multiple of y^n, so that

$$\sum_{i=n}^{n+k} a_i x^i = \lambda y^n \ .$$

Restricting to S^{2n} we see that $\lambda = 1$, and so

$$y^n = (\log(1 + x))^n$$

has all coefficients from x^n to x^{n+k} <u>integral</u> as required.

<u>Remark:</u> It has been shown by Adams and Grant-Walker (Proc. Camb. Phil. Soc. 61(1965), 81-103) that (3. 2. 10) gives a <u>sufficient</u> condition for the existence of a retraction.

Suppose once more that we have a map $f : S^{2m+2n-1} \to S^{2m}$. Then we can attach to f an invariant $e(f) \in Q/Z$ in the following fashion.

Let X be the mapping cone of f, $i = S^{2m} \to X$ the inclusion, $j : X \to S^{2n+2m}$ the map which collapses S^{2m}. Let u generate $\widetilde{K}^0(S^{2n+2m})$, v generate $\widetilde{K}^0(S^{2m})$, and let $x \in \widetilde{K}^0(X)$ be such that $i^*(x) = v$. Let $y = j^*(u)$. Then for any k,

$$\psi^k(x) = k^m x + a_k y \quad .$$

As before, we know that $\psi^k\psi^\ell = \psi^\ell\psi^k$, so that

$$k^n(k^m - 1)a_\ell = \ell^n(\ell^m - 1)a_k .$$

Thus

$$e(f) = \frac{a_k}{k^n(k^m - 1)} \in Q$$

is well defined once x is chosen. If x is changed by a multiple of y, $e(f)$ is changed by an integer, so that $e(f) \in Q/Z$ is well defined. We leave to the reader the elementary exercise that $e : \pi_{2n+2m-1}(S^{2m}) \to Q/Z$ is a group homomorphism. It turns out that this is a very powerful invariant.

§3. The Groups $J(X)$. In this section we assume, for simplicity, that X is connected. One can introduce a notion of equivalence between vector bundles, known as fibre homotopy equivalence, which is of much interest in homotopy theory. Let E, E' be two bundles over a space X, and suppose that both E, E' have been given Hermitian metrics. Then E and E' are said to be fibre homotopy equivalent if there exist maps $f : S(E) \to S(E')$, $g : S(E') \to S(E)$, commuting with the projection onto X, and such that gf and fg are homotopic to the identity through fibre-preserving maps. Clearly this is an equivalence relation defined on the set of equivalence classes of vector bundles over X.

Fibre homotopy equivalence is additive; that is, if E, E' are fibre homotopy equivalent to F, F' respectively, then $E \oplus E'$ is fibre-homotopy equivalent to $F \oplus F'$. This follows from the fact that $S(E \oplus E')$ may be viewed as the fibre-join of the two fibre spaces $S(E), S(E')$: in general the fibre-join of $\pi : Y \to X$, $\pi' : Y' \to X$ is defined as the space of triples (y, t, y') where $t \in I$, $\pi(y) = \pi'(y')$ and we impose the equivalence relations

$$(y, 0, y_1') \sim (y, 0, y_2')$$

$$(y_2, 1, y') \sim (y_2, 1, y') \quad .$$

We say that two bundles E, E' are stably fibre-homotopy equivalent if there exist trivial bundles V, V' such that $E \oplus V$ is fibre-homotopy equivalent to $E' \oplus V'$. The set of all stable fibre-homotopy equivalence classes over X forms a semi-group which we denote by $J(X)$. Since every vector bundle E has a complementary bundle F so that $E \oplus F$ is trivial it follows that $J(X)$ is a group and hence the map

$$\mathrm{Vect}(X) \longrightarrow J(X)$$

extends to an epimorphism

$$K(X) \longrightarrow J(X)$$

which we also denote by J.

If we have two bundles E, E' and if $\pi : S(E) \to X$, $\pi' : S(E') \to X$ are the projection maps of the respective sphere bundles, the Thom complexes X^E, $X^{E'}$ are just the mapping cones of the maps π, π' respectively. Thus, we see that if E and E' are fibre homotopy equivalent, X^E and $X^{E'}$ have the same homotopy type. However, if E is a trivial bundle of dimension n, $X^E = S^{2n}(X^+)$. Thus, to show that $J(E) \neq 0$, it suffices to show that X^E does not have the same stable homotopy type as a suspension of X^+.

We shall now show how to use the operations ψ^k of §2 to give necessary conditions for $J(E) = 0$. By the Thom isomorphism (2.7.12) we know that $\tilde{K}(X^E)$ is a free $K(X)$-module generated by

λ_E. Hence, for any k, there is a unique element $\rho^k(E) \in K(X)$ such that

$$\psi^k(\lambda_E) = \lambda_E \, \rho^k(E) \quad .$$

The multiplicative property of the fundamental class λ_E, established in §2, together with the fact that ψ^k preserves products, shows that

$$\rho^k(E \oplus E') = \rho^k(E) \cdot \rho^k(E') \quad .$$

Also, taking $E = 1$, and recalling that

$$\psi^k \circ \beta = k\beta \circ \psi^k$$

where β is the periodicity isomorphism, we see that

$$\rho^k(1) = k \quad .$$

Now let $Q_k = Z[1/k]$ be the subring of Q consisting of fractions with denominators a power of k. Then if we put

$$\sigma^k(E) = k^{-n} \rho_k(E) \qquad\qquad n = \dim E$$

we obtain a homomorphism

$$\sigma^k : K(X) \to G_k$$

where G_k is the <u>multiplicative group of units</u> of $K(X) \otimes Q_k$. Suppose now E is fibre-homotopically trivial, then there exists

$u \in \widetilde{K}(X^E)$ so that $\psi^k u = k^n u$. Putting $u = \lambda_E\, a$ we find that

$$\psi^k \lambda_E \cdot \psi^k a = k^n \lambda_E\, a$$

and so

$$\sigma^k(E) \cdot \psi^k(a) = a \quad .$$

Moreover, restricting to a point, we see that a has augmentation 1 so that a and $\psi^k(a)$ are both elements of G_k. Hence we may write

$$\sigma^k(E) = \frac{a}{\psi^k(a)} \in G_k \quad .$$

Since $\sigma^k(E)$ depends only on the stable class of E, we have established the following

PROPOSITION 3.3.1. Let $H_k \subset G_k$ be the subgroup generated by all elements of the form $a/\psi^k(a)$ with a a unit of $K(X)$. Then

$$\sigma^k : K(X) \longrightarrow G_k$$

maps the kernel of J into H_k, and so induces a homomorphism

$$J(X) \longrightarrow G_k/H_k \quad .$$

In order to apply (3. 3. 1) it is necessary to be able to compute σ^k or equivalently ρ^k. Now

$$\rho^k \in \mathrm{Op}\, K$$

is an operation. Its augmentation is known so it remains to determine its value on combinations of line-bundles. Because of its multiplicative property, it is only necessary to determine $\rho^k(L)$ for a line-bundle L.

LEMMA 3. 3. 2. <u>For a line-bundle</u> L, <u>we have</u>

$$\rho^k[L] = \sum_{j=0}^{k-1} [L]^j .$$

<u>Proof</u>: By (2. 7. 1) and (2. 7. 2) we have a description of $\tilde{K}(X^L)$ as the $K(X)$ sub-module of $K(P(L \oplus 1))$ generated by $n = 1 - [L][H]$. The structure of $K(P(L \oplus 1))$ is of course given by our main theorem (2. 2. 1). Hence

$$\begin{aligned}\psi^k(u) &= 1 - [L^k][H^k] \\ &= (1 - [L][H]) \left\{ \sum_{j=0}^{k-1} [L^j][H^j] \right\} \\ &= u \sum_{j=0}^{k-1} [L^j], \quad \text{since } (1 - [L][H])(1 - [H]) = 0 .\end{aligned}$$

Thus

$$\psi^k \lambda_L = \lambda_L \left\{ \sum_{j=0}^{k-1} [L^j] \right\}$$

proving that

$$\rho^k(L) = \sum_{j=0}^{k-1} [L^j]$$

as required.

As an example we take $X = P_{2n}(R)$, real projective 2n-space. As shown in (2. 7. 7) $\tilde{K}(X)$ is cyclic of order 2^n with generator $x = [L] - 1$, where L is the standard line-bundle. The multiplicative structure follows from the relation $[L]^2 = 1$ (since L is associated to the group Z_2). Now take $k = 3$, then

$$\psi^3(x) = [L^3] - 1 = x ,$$

and so the group H_3 defined above is reduced to the identity. Using (3. 3. 2) we find

$$\begin{aligned}
\sigma^3(mx) &= \rho^3(mx) = (\rho^3(x))^m = (\rho^3[L])^m \cdot 3^{-m} \\
&= 3^{-m}(1 + [L] + [L]^2)^m \\
&= (1 + x/3)^m \\
&= 1 + \sum_{i=1}^{m} (-1)^{i-1} \frac{2^{i-1}}{3^i} \binom{m}{i} x \qquad \text{(since } x^2 = -2x) \\
&= 1 + \frac{1}{2}\left(1 - \left(1 - \frac{2}{3}\right)^m\right) x \\
&= 1 + 3^{-m}\left(\frac{3^m - 1}{2}\right) x \qquad .
\end{aligned}$$

Thus if $J(mx) = 0$ we must have $3^m - 1$ divisible by 2^{n+1}. This happens if and only if 2^{n-1} divides m. Thus the kernel of

$$J : \widetilde{K}(P_{2n}(R)) \to J(P_{2n}(R))$$

is at most of order 2. This result can in fact be improved by use of real K-theory and is the basis of the solution of the vector-field problem for spheres.

The problem considered in (3.2.10) is in fact a special case of the more general problem we are considering now. In fact, the space $P_{n+k}(C)/P_{n-1}(C)$ is easily seen to be the Thom space of the bundle nH over $P_k(C)$. The conclusion of (3.2.10) may therefore be interpreted as a statement about the order of $J[H] \in J(P_k(C))$. The method of proof in (3.2.10) is essentially the same as that used in this section. The point is that we are now considering not just a single space but a whole class, namely Thom spaces, and describing a uniform method for dealing with all spaces of this class.

For further details of $J(X)$ on the preceding lines we refer the reader to the series of papers "On the groups $J(X)$" by J. F. Adams (Topology 1964-).

APPENDIX

The space of Fredholm operators. In this appendix we shall give a Hilbert space interpretation[†] of $K(X)$. This is of interest in connection with the theory of the index for elliptic operators.

Let H denote a separable complex Hilbert space, and let $\mathcal{A}(H)$ be the algebra of all bounded operators on H. We give $\mathcal{A}$ the norm topology. It is well-known that this makes $\mathcal{A}$ into a Banach algebra. In particular the group of units $\mathcal{A}^*$ of $\mathcal{A}$ forms an open set. We recall also that, by the closed graph theorem, any $T \in \mathcal{A}$ which is an algebraic isomorphism $H \to H$ is also a topological isomorphism, i.e., T^{-1} exists in $\mathcal{A}$ and so $T \in \mathcal{A}^*$.

DEFINITION: An operator $T \in \mathcal{A}(H)$ is a Fredholm operator if Ker T and Coker T are finite dimensional. The integer

$$\dim \operatorname{Ker} T - \dim \operatorname{Coker} T$$

is called the index of T.

We first observe that, for a Fredholm operator T, the image $T(H)$ is closed. In fact, since $T(H)$ is of finite codimension in H we can find a finite dimensional algebraic complement P. Then $T \oplus j : H \oplus P \to H$ (where $j : P \to H$ is the inclusion) is

[†] These results have been obtained independently by K. Janich (Bonn dissertation 1964).

surjective, and so by the closed graph theorem the image of any closed set is closed. In particular $T(H) = T \oplus j(H \oplus 0)$ is closed.

Let $\mathfrak{F} \subset \mathfrak{G}$ be the subspace of all Fredholm operators. If T, S are two Fredholm operators we have

$$\dim \operatorname{Ker} TS \leq \dim \operatorname{Ker} T + \dim \operatorname{Ker} S$$
$$\dim \operatorname{Coker} TS \leq \dim \operatorname{Coker} T + \dim \operatorname{Coker} S$$

and so TS is again a Fredholm operator. Thus $\mathfrak{F}$ is a topological space with an associative product $\mathfrak{F} \times \mathfrak{F} \to \mathfrak{F}$. Hence for any space X the set $[X, \mathfrak{F}]$ of homotopy classes of mappings $X \to \mathfrak{F}$ is a semi-group. Our main aim will be to indicate the proof of the following:

THEOREM A1. <u>For any compact space we have a natural isomorphism</u>

$$\text{index} : [X, \mathfrak{F}] \to K(X) \ .$$

<u>Note</u>: If X is a point this means that the connected components of $\mathfrak{F}$ are determined by an integer: this is in fact the index which explains our use of the word in the more general context of Theorem A1 .

Theorem A1 asserts that $\mathfrak{F}$ is a classifying or representing space for K-theory. Another closely related classifying space may be obtained as follows. Let $\mathcal{K} \subset \mathcal{Q}$ denote all the *compact* operators. This is a closed 2-sided ideal and the quotient $\mathcal{B} = \mathcal{Q}/\mathcal{K}$ is therefore again a Banach algebra. Let $\mathcal{B}^*$ be the group of units of $\mathcal{B}$. It is a topological group and so, for any X, $[X, \mathcal{B}^*]$ is a group. Then our second theorem is:

THEOREM A2. $\mathcal{B}^*$ *is a classifying space for K-theory, i.e., we have a natural group-isomorphism*

$$[X, \mathcal{B}^*] \cong K(X) \quad .$$

We begin with the following lemma which is essentially the generalization to infinite dimensions of Proposition 1.3.2.

LEMMA A3. *Let* $T \in \mathfrak{F}$ *and let* V *be a closed subspace of* H *of finite codimension such that* $V \cap \operatorname{Ker} T = 0$. *Then there exists a neighborhood* U *of* T *in* $\mathcal{Q}$ *such that, for all* $S \in U$, *we have*

(i) $V \cap \operatorname{Ker} S = 0$

(ii) $\bigcup_{S \in U} H/S(V)$ *topologized as a quotient space of* $U \times H$ *is a trivial vector bundle over* U.

Proof: Let $W = T(V)^{\perp}$ (the orthogonal complement of $T(V)$ in H.) Since $T \in \mathfrak{F}$ and $\dim H/V$ is finite it follows that $\dim W$ is finite. Now define, for $S \in \mathfrak{a}$,

$$\varphi_S : V \oplus W \to H$$

by $\varphi_S(V \oplus W) = S(V) + W$. Then $S \to \varphi_S$ gives a continuous linear map

$$\varphi : \mathfrak{a} \to \mathfrak{L}(V \oplus W, H)$$

where $\mathfrak{L}$ stands for the space of all continuous linear maps with the norm topology. Now φ_T is an isomorphism and the isomorphisms in $\mathfrak{L}$ form an open set (like $\mathfrak{a}^*$ in $\mathfrak{a}$). Hence there exists a neighborhood U of T in $\mathfrak{a}$ so that φ_S is an isomorphism for all $S \in U$. This clearly implies (i) and (ii).

COROLLARY A4. $\mathfrak{F}$ is open in $\mathfrak{a}$.

Proof: Take $V = (\operatorname{Ker} T)^{\perp}$ in (A3).

PROPOSITION A5. Let $T : X \to \mathfrak{F}$ be a continuous map with X compact. Then there exists $V \subset H$, closed and of finite codimension so that

(i) $V \cap \operatorname{Ker} T_x = 0$ for all $x \in X$.

Moreover, for any such V we have

(ii) $\bigcup_{x \in X} H/T_x(V)$, topologized as a quotient space of $X \times H$, is a vector bundle over X.

<u>Proof:</u> For each $x \in X$ take $V_x = (\mathrm{Ker}\, T_x)^{\perp}$ and let U_x be the inverse image under T of the open set given by (A3). Let $K_i = U_{x_i}$ be a finite sub-cover of this family of open sets. Then $V = \cap_i V_{x_i}$ satisfies (i). To prove (ii) we apply (A3) to each T_x, and deduce that $\cup_y H/T_y(V)$ is locally trivial near x, and hence is a vector bundle.

For brevity we shall denote the bundle $\cup_{x \in X} H/T_x(V)$, occurring in (A4), by $H/T(V)$. Just as in the finite-dimensional case we can split the map $p : X \times H \to H/T(V)$; more precisely we can find a continuous map

$$\varphi : H/T(V) \to X \times H$$

commuting with projection on X and such that

$$p\varphi = \text{identity}$$

One way to construct φ is to use the metric in H and map $H/T(V)$ onto the orthogonal complement $T(V)^{\perp}$ of $T(V)$. This is technically inconvenient since we then have to verify that $T(V)^{\perp}$ is a vector bundle. Instead we observe that, by definition, p splits locally and so we can choose splittings φ_i over U_i, where U_i is a finite open covering of X. Then $\varphi_i - \varphi_j = \theta_{ij}$ is essentially a map $H/T(V)|U_i \cap U_j \to U_i \cap U_j \times V$. If ρ_i is a partition of unity subordinate to the covering we put, in the usual way

$$\theta_i = \Sigma \rho_j \theta_{ij}$$

so that θ_i is defined over all U_i, and then $\varphi = \varphi_i - \theta_i$ is independent of i and gives the required splitting.

We can now define index T for any map $T : X \to \mathfrak{F}$ (X being compact). We choose V as in (A5) and put

$$\text{index } T = [H/V] - [H/T(V)] \in K(X) ,$$

where H/V stands for the trivial bundle $X \times H/V$. We must show that this is independent of the choice of V. If W is another choice so is $V \cap W$, so it is sufficient to assume $W \subset V$. But then we have the exact sequences of vector bundles

$$0 \longrightarrow V/W \longrightarrow H/W \longrightarrow H/V \longrightarrow 0$$

$$0 \longrightarrow V/W \longrightarrow H/T(W) \longrightarrow H/T(V) \longrightarrow 0 .$$

Hence

$$[H/V] - [H/W] = [V/W] = [H/T(V)] - [H/T(W)]$$

as required.

It is clear that our definition of index T is functorial. Thus if $f : Y \to X$ is a continuous map then

$$\text{index } Tf = f^* \text{ index } T .$$

This follows from the fact that a choice of the subspace V for T is also a choice for Tf.

If $T : X \times I \to \mathfrak{F}$ is a homotopy between T_0 and T_1 then index $T \in K(X \times I)$ restricts to index $T_i \in K(X \times \{i\})$, $i = 0, 1$. Since we know that

$$K(X \times I) \twoheadrightarrow K(X \times \{i\}) \cong K(X)$$

is an isomorphism, it follows that

$$\text{index } T_0 = \text{index } T_1 .$$

Thus

$$\text{index} : [X, \mathfrak{F}] \longrightarrow K(X)$$

is well-defined.

Next we must show that "index" is a homomorphism. Let $S : X \to \mathfrak{F}$, $T : X \to \mathfrak{F}$ be two continuous maps. Let $W \subset H$ be a choice for T. Replacing S by the homotopic map $\pi_W S$ (π_W denoting projection onto W) we can assume $S(H) \subset W$. Now let $V \subset H$ be a choice for S then it is also a choice for TS and we have an exact sequence of vector bundles over X

$$0 \longrightarrow W/SV \xrightarrow{T} H/TSV \longrightarrow H/TW \longrightarrow 0 .$$

Hence

$$\begin{aligned} \text{index } TS &= [H/V] - [H/TSV] \\ &= [H/V] - [W/SV] - [H/TW] \\ &= [H/V] - [H/SV] + [H/W] - [H/TW] \\ &= \text{index } S + \text{index } T \end{aligned}$$

as required.

Having now established that

$$\text{index} : [X, \mathfrak{F}] \longrightarrow K(X)$$

is a homomorphism the next step in the proof of Theorem (A1) is

PROPOSITION A6. <u>We have an exact sequence of semi-groups</u>

$$[X, \mathcal{G}^*] \longrightarrow [X, \mathfrak{F}] \xrightarrow{\text{index}} K(X) \longrightarrow 0 \ .$$

<u>Proof:</u> Consider first a map $T : X \to \mathfrak{F}$ of index zero. This means that

$$[H/V] - [H/TV] = 0 \qquad \text{in } K(X) \ .$$

Hence adding a trivial bundle P to both factors we have

$$H/V \oplus P \cong H/TV \oplus P \, .$$

Equivalently replacing V by a closed subspace W with $\dim V/W = \dim P$,

$$H/W \cong H/TW \quad .$$

If we now split $X \times H \to H/TW$ as explained earlier we obtain a continuous map

$$\varphi : X \times H/W \longrightarrow X \times H$$

commuting with projection on X, linear on the fibres. If

$$\Phi : X \longrightarrow \mathcal{L}(H/W, H)$$

is the map associated to φ, it follows from the construction of φ that

$$x \longrightarrow \Phi_x + T_x$$

gives a continuous map

$$X \longrightarrow \mathcal{A}^* \quad .$$

But if $0 \leq t \leq 1$, $T + t\Phi$ provides a homotopy of maps $X \to \mathfrak{F}$ connecting T with $T + \Phi$. This proves exactness in the middle.

It remains to show that the index is surjective. Let E be a vector bundle over X and let F be a complement so that $E \oplus F$ is isomorphic to the trivial bundle $X \times V$. Let $\pi_x \in \operatorname{End} V$

denote projection onto the subspace corresponding to E_x.
Let $T_k \in \mathfrak{F}$ denote the standard operator of index k, defined relative to an orthonormal basis $\{e_i\}$ $(i = 1, 2, \ldots)$ by

$$\begin{aligned} T_k(e_i) &= e_{i-k} && \text{if } i - k \geq 1 \\ &= 0 && \text{otherwise} . \end{aligned}$$

Then define a map

$$S : X \to \mathfrak{F}(H \otimes V) \cong \mathfrak{F}(H)$$

by $S_x = T_{-1} \otimes \pi_x + T_0 \otimes (1 - \pi_x)$. We have $\operatorname{Ker} S_x = 0$ for all x, and $H \otimes V / S(H \otimes V) \cong E$. Hence

$$\text{index } S = -[E] .$$

The constant map $T_k : X \to \mathfrak{F}$ given by $T_k(x) = T_k$ has index k and so

$$\text{index } T_k S = k - [E] .$$

Since every element of $K(X)$ is of the form $k - [E]$ this shows that the index is surjective and completes the proof of the proposition.

Theorem (A1) now follows from (A6) and the following:

PROPOSITION A7. $[X, \mathfrak{a}^*] = 1$.

This proposition is due to Kuiper and we shall not reproduce the proof here (full details are in Kuiper's paper: Topology 3 (1964) 19-30). In fact, Kuiper actually shows that $\mathcal{G}^*$ is contractible.

We turn now to discuss the proof of (A2). We recall first that

$$1 + \mathcal{K} \subset \mathfrak{F} \quad .$$

This is a standard result in the theory of compact operators: the proof is easy.

PROPOSITION A8. Let $\pi : \mathcal{G} \to \mathcal{B} = \mathcal{G}/\mathcal{K}$ be the natural map. Then

$$\mathfrak{F} = \pi^{-1}(\mathcal{B}^*) \quad .$$

Proof: (a) Let $T \in \mathfrak{F}$ and let P, Q denote orthogonal projection onto Ker T, Ker T^* respectively. Then $T^*T + P$ and $TT^* + Q$ are both in $\mathcal{G}^*$, and so their images by π are in $\mathcal{B}^*$. But $P, Q \in \mathcal{K}$ and so $\pi(T^*) \cdot \pi(T) \in \mathcal{B}^*$, $\pi(T)\pi(T^*) \in \mathcal{B}^*$. This implies that $\pi(T) \in \mathcal{B}^*$.

(b) Let $T \in \pi^{-1}(\mathcal{B}^*)$, i.e., there exists $S \in \mathcal{G}$ with ST and $TS \in 1 + \mathcal{K} \subset \mathfrak{F}$. Since dim Ker T $\leq$ dim Ker ST

$$\dim \text{Coker } T \leq \dim \text{Coker } TS$$

it follows that $T \in \mathfrak{F}$.

Theorem (A2) will now follow from (A1) and the following general lemma (applied with $L = \mathcal{G}$, $M = \mathcal{B}$, $U = \mathcal{B}^*$).

LEMMA A9. Let $\pi : L \to M$ be a continuous linear map of Banach spaces with $\pi(L)$ dense in M and let U be an open set in M. Then, for any compact X

$$[X, \pi^{-1}(U)] \longrightarrow [X, U]$$

is bijective.

Proof: First we shall show that if

$$\pi : L \longrightarrow M$$

satisfies the hypotheses of the lemma, then for any compact X, the induced map

$$\pi^X : L^X \longrightarrow M^X$$

also satisfies the same hypotheses. Since L^X, M^X are Banach spaces the only thing to prove is that $\pi^X(L^X)$ is dense in M^X. Thus, let $f : X \to M$ be given. We have to construct $g : X \to L$ so that $\|\pi g(x) - f(x)\| < \epsilon$ for all $x \in X$. Choose $a_1, \cdots, a_n$ in $f(X)$ so that their $\frac{\epsilon}{3}$-neighborhoods $\{U_i\}$ cover $f(X)$ and choose b_i so that $\|\pi(b_i) - a_i\| < \epsilon/3$. Let $u_i(x)$ be a partition of unity of X subordinate to the covering $\{f^{-1}U_i\}$ and define

$g : X \to L$ by

$$g(x) = \sum u_i(x)\, b_i \quad .$$

This is the required map.

Hence replacing π by π^X and U by U^X (which is open in M^X) we see that it is only necessary to prove the lemma when X is a point, i.e., to prove that

$$\pi^{-1}(U) \longrightarrow U$$

induces a bijection of path-components. Clearly this map of path-components is surjective: if $P \in U$ then there exists $Q \in \pi(L) \cap U$ such that the segment PQ is entirely in U. To see that it is injective let $P_0, P_1 \in \pi^{-1}(U)$ and suppose $f : I \to U$ is a path with $f(0) = \pi(P_0)$, $f(1) = \pi(P_1)$. By what we proved at the beginning there exists $g : I \to \pi^{-1}(U)$ such that

$$\| \pi g(t) - f(t) \| < \epsilon \qquad\qquad \text{for all } t \in I \ .$$

If ϵ is sufficiently small the segments joining $\pi g(i)$ to $f(i)$, for $i = 0, 1$, will lie entirely in U. This implies that the segment joining $g(i)$ to P_i, for $i = 0, 1$, lies in $\pi^{-1}(U)$. Thus P_0 can be joined to P_1 by a path in $\pi^{-1}(U)$ (see figure) and this completes the proof.

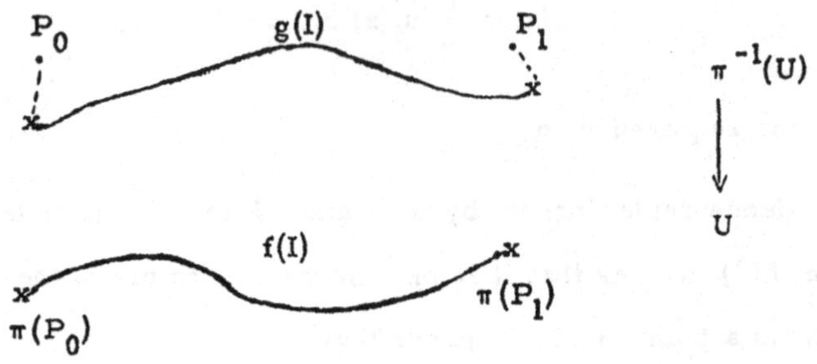

References

[1] R. Bott, The stable homotopy of classical groups, Ann. of Math. 70 (1959) 313-337.

[2] M.F. Atiyah, Bott periodicity and the index of elliptic operators, Quart. J. Math. Oxford (2) 19 (1968) 113-140.

[3] M.F. Atiyah and R.Bott, On the periodicity theorem for complex vector bundles, Acta. Math. 112 (1964) 299-347.

[4] M.F. Atiyah,Algebraic topology and operators in Hilbert space, Springer Lecture Notes in Mathematics 103 (1969) 101-122.

Reprinted from Quart. J. Math. (Clarendon Press, Oxford)

POWER OPERATIONS IN K-THEORY

By M. F. ATIYAH (*Oxford*)

[Received 10 January 1966]

Introduction

FOR any finite CW-complex X we can define the Grothendieck group $K(X)$. It is constructed from the set of complex vector bundles over X [see (**8**) for precise definitions]. It has many formal similarities to the cohomology of X, but there is one striking difference. Whereas cohomology is *graded*, by dimension, $K(X)$ has only a *filtration*: the subgroup $K_q(X)$ is defined as the kernel of the restriction homomorphism

$$K(X) \to K(X_{q-1}),$$

where X_{q-1} is the $(q-1)$-skeleton of X. Now $K(X)$ has a ring structure, induced by the tensor product of vector bundles, and this is compatible with the filtration, so that $K(X)$ becomes a filtered ring. There are also natural operations in $K(X)$, induced by the exterior powers, and one of the main purposes of this paper is to examine the relation between operations and filtration (Theorem 4.3).

Besides the formal analogy between $K(X)$ and cohomology there is a more precise relationship. If X has no torsion this takes a particularly simple form, namely the even-dimensional part of the integral cohomology ring

$$H^{\mathrm{ev}}(X; \mathbf{Z}) = \sum_q H^{2q}(X; \mathbf{Z})$$

is naturally isomorphic to the graded ring

$$GK(X) = \sum_q K_{2q}(X)/K_{2q-1}(X).$$

Since this isomorphism preserves the ring structures, it is natural to ask about the operations. Can we relate the operations in K-theory to the Steenrod operations in cohomology?

If we consider the way the operations arise in the two theories, we see that in both cases a key role is played by the symmetric group. It is well known [cf. (**10**)] that one way of introducing the Steenrod operations is via the cohomology of the symmetric group (and its subgroups). On the other hand, the operations on vector bundles come essentially from representations of the general linear group and the role of the symmetric group in constructing the irreducible representations of $GL(n)$ is of course classical [cf. (**11**)]. A closer examination of the two cases shows that the symmetric group enters in essentially the same way in both theories. The operations arise from the interplay of the kth power map and the action of the symmetric group S_k.

We shall develop this point of view and, following Steenrod, we shall introduce operations in K-theory corresponding to any subgroup G of S_k. Taking $k = p$ (a prime) and $G = Z_p$ to be the cyclic group of order p we find that the only non-trivial operation defined by Z_p is the Adams operation ψ^p. This shows that ψ^p is analogous to the total Steenrod power operation $\sum P^i$ and, for spaces without torsion, we obtain the precise relationship between ψ^p and the P^i (Theorem 6.5). Incidentally we give a rather simple geometrical description (2.7) of the operation ψ^p.

It is not difficult to translate Theorem 6.5 into rational cohomology by use of the Chern character, and (for spaces without torsion) we recover a theorem of Adams (**1**). In fact this paper originated in an attempt to obtain Adams's results by more direct and elementary methods.

Although the only essentially new results are concerned with the relation between operations and filtration, it seems appropriate to give a new self-contained account of the theory of operations in K-theory. We assume known the standard facts about K-theory [cf. (**8**)] and the theory of representations of *finite* groups. We do not assume anything about representations of compact Lie groups.

In § 1 we present what is relevant from the classical theory of the symmetric group and tensor products. We follow essentially an idea of Schur [see (**11**) 215], which puts the emphasis on the symmetric group S_k rather than the general linear group $GL(n)$. This seems particularly appropriate for K-theory where the dimension n is rather a nuisance (it can even be negative!). Thus we introduce a graded ring

$$R_* = \sum_k \mathrm{Hom}_{\mathbf{Z}}(R(S_k), \mathbf{Z}),$$

where $R(S_k)$ is the character ring of S_k, and we study this in considerable detail. Among the formulae we obtain, at least one (Proposition 1.9) is probably not well known. In § 2, by considering the tensor powers of a graded vector bundle, we show how to define a ring homomorphism

$$j: R_* \to \mathrm{Op}(K),$$

where $\mathrm{Op}(K)$ stands for the operations in K-theory. The detailed information about R_* obtained in § 1 is then applied to yield results in K-theory.

§ 3 is concerned with 'externalizing' and 'relativizing' the tensor powers defined in § 2. Then in § 4 we study the relation of operations and filtration. § 5 is devoted to the cyclic group of prime order and its related operations. In § 6 we investigate briefly our operations in connexion with the spectral sequence $H^*(X, \mathbf{Z}) \Rightarrow K^*(X)$ and obtain in particular the relation with the Steenrod powers mentioned earlier. Finally in § 7 we translate things into rational cohomology and derive Adams's result.

The general exposition is considerably simplified by introducing the functor $K_G(X)$ for a G-space X (§ 2). We establish some of its elementary properties but for a fuller treatment we refer to (**4**) and (**9**).

The key idea that one should consider the symmetric group acting on the kth power of a complex of vector bundles is due originally to Grothendieck, and there is a considerable overlap between our presentation of operations in K-theory and some of his unpublished work.

I am indebted to P. Cartier and B. Kostant for some very enlightening discussions.

1. Tensor products and the symmetric group

For any finite group G we denote by $R(G)$ the free abelian group generated by the (isomorphism classes of) irreducible complex representations of G. It is a ring with respect to the tensor product. By assigning to each irreducible representation its character we obtain an embedding of $R(G)$ in the ring of all complex-valued class functions on G. We shall frequently identify $R(G)$ with this subring and refer to it as the *character ring* of G. For any two finite groups G, H we have a natural isomorphism

$$R(G) \otimes R(H) \to R(G \times H).$$

Now let S_k be the symmetric group and let $\{V_\pi\}$ be a complete set of irreducible complex S_k-modules. Here π may be regarded as a partition of k, but no use will be made of this fact. Let E be a complex vector space, $E^{\otimes k}$ its kth tensor power. The group S_k acts on this in a natural way, and we consider the classical decomposition

$$E^{\otimes k} \cong \sum V_\pi \otimes \pi(E),$$

where $\pi(E) = \mathrm{Hom}_{S_k}(V_\pi, E^{\otimes k})$. We note in particular the two extreme cases: if V_π is the trivial one-dimensional representation, then $\pi(E)$ is the kth symmetric power $\sigma^k(E)$; if V_π is the sign representation, then $\pi(E)$ is the kth exterior power $\lambda^k(E)$. Any endomorphism T of E induces an S_k-endomorphism $T^{\otimes k}$ of $E^{\otimes k}$, and hence an endomorphism $\pi(T)$ of $\pi(E)$. Taking $T \in GL(E)$, we see that $\pi(E)$ becomes a representation space of $GL(E)$, and this is of course the classical construction for the irreducible representations of the general linear group. For our purposes, however, this is not relevant. All we are interested in are the character formulae. We therefore proceed as follows.

Let $E = \mathbf{C}^n$ and let T be the diagonal matrix $(t_1, \ldots, t_n)$. Since the eigenvalues of $T^{\otimes k}$ are all monomials of degree k in $t_1, \ldots, t_n$, it follows that, for each π, $\operatorname{Trace} \pi(T)$ is a homogeneous polynomial in $t_1, \ldots, t_n$ with integer coefficients. Moreover, $\operatorname{Trace} \pi(T) = \operatorname{Trace}(\pi(S^{-1}TS))$ for any permutation matrix S and so $\operatorname{Trace} \pi(T)$ is symmetric in $t_1, \ldots, t_n$. We define

$$\Delta_{n,k} = \operatorname{Trace}_{S_k}(T^{\otimes k}) = \sum_{\pi} \operatorname{Trace} \pi(T) \otimes [V_\pi] \in \operatorname{Sym}_k[t_1, \ldots, t_n] \otimes R(S_k),$$

where $[V_\pi] \in R(S_k)$ is the class of V_π and $\operatorname{Sym}_k[t_1, \ldots, t_n]$ denotes the symmetric polynomials of degree k. If we regard $R(S_k)$ as the character ring, then $\Delta_{n,k}$ is just the function of $t_1, \ldots, t_n$ and $g \in S_k$ given by $\operatorname{Trace}(gT^{\otimes k})$. There are a number of other ways of writing this basic element, the simplest being the following proposition:

PROPOSITION 1.1. *For any partition* $\alpha = (\alpha_1, \ldots, \alpha_r)$ *of* k *let* $\rho_\alpha \in R(S_k)$ *be the representation induced from the trivial representation of*

$$S_\alpha = S_{\alpha_1} \times S_{\alpha_2} \times \ldots \times S_{\alpha_r},$$

then

$$\Delta = \sum_{\alpha \vdash k} m_\alpha \otimes \rho_\alpha,$$

where m_α *is the monomial symmetric function generated by* $t_1^{\alpha_1} t_2^{\alpha_2} \ldots t_r^{\alpha_r}$ *and the summation is over all partitions of* k.

Proof. Let E^α be the eigenspace of $T^{\otimes k}$ corresponding to the eigenvalue $t_1^{\alpha_1} t_2^{\alpha_2} \ldots t_r^{\alpha_r}$. This has as a basis the orbit under S_k of the vector

$$e_\alpha = e_1^{\otimes \alpha_1} \otimes e_2^{\otimes \alpha_2} \ldots \otimes e_r^{\otimes \alpha_r},$$

where $e_1, \ldots, e_n$ are the standard base of $\mathbf{C}^n$. Since the stabilizer of e_α is just the subgroup S_α, it follows that E^α is the induced representation ρ_α. Since S_α and S_β are conjugate if α and β are the same *partition* of k, it follows that

$$\Delta = \sum_{|\alpha| = k} t^\alpha \otimes \rho_\alpha = \sum_{\alpha \vdash k} m_\alpha \otimes \rho_\alpha,$$

where the first summation is over all *sequences* $\alpha_1, \alpha_2, \ldots$ with

$$|\alpha| = \sum \alpha_i = k.$$

Now let us introduce the dual group

$$R_*(S_k) = \operatorname{Hom}_{\mathbf{Z}}(R(S_k), \mathbf{Z}).$$

Then $\Delta_{n,k}$ defines (and is defined by) a homomorphism

$$\Delta'_{n,k} \colon R_*(S_k) \to \operatorname{Sym}_k[t_1, \ldots, t_n].$$

From the inclusions $\qquad S_k \times S_l \to S_{k+l}$

we obtain homomorphisms

$$R(S_{k+l}) \to R(S_k \times S_l) \cong R(S_k) \otimes R(S_l)$$

and hence by duality

$$R_*(S_k) \otimes R_*(S_l) \to R_*(S_{k+l}).$$

Putting $R_* = \sum_{k \geqslant 0} R_*(S_k)$ we see that the above pairings turn R_* into a *commutative graded ring*. This follows from the fact, already used in Proposition 1.1, that S_α and S_β are conjugate if α and β are the same partition. Moreover, if we define

$$\Delta'_n : R_* \to \mathrm{Sym}[t_1, \ldots, t_n]$$

by $\Delta'_n = \sum \Delta'_{n,k}$, we see that Δ'_n is a *ring homomorphism*. This follows from the multiplicative property of the trace:

$$\mathrm{Trace}(g_1 g_2 T^{\otimes(k+l)}) = \mathrm{Trace}(g_1 T^{\otimes k})\mathrm{Trace}(g_2 T^{\otimes l}),$$

where $g_1 \in S_k$, $g_2 \in S_l$. Finally we observe that we have a commutative diagram

$$\begin{array}{ccc} R_* & \xrightarrow{\quad \Delta'_{n+1} \quad} & \mathrm{Sym}[t_1, \ldots, t_n] \\ & \searrow^{\Delta'_n} & \downarrow \\ & & \mathrm{Sym}[t_1, \ldots, t_n] \end{array}$$

where the vertical arrow is given by putting $t_{n+1} = 0$. Hence passing to the limit we can define

$$\Delta' : R_* \to \lim_{\substack{\leftarrow \\ n}} \mathrm{Sym}[t_1, \ldots, t_n].$$

Here the inverse limit is taken in the category of *graded* rings, so that

$$\lim_{\substack{\leftarrow \\ n}} \mathrm{Sym}[t_1, \ldots, t_n] = \sum_{k=0} \lim_{\substack{\leftarrow \\ n}} \mathrm{Sym}_k[t_1, \ldots, t_n]$$

is the direct sum (and not the direct product) of its homogeneous parts.

Proposition 1.2. $\Delta' : R_* \to \lim_{\substack{\leftarrow \\ n}} \mathrm{Sym}[t_1, \ldots, t_n]$ *is an isomorphism.*

Proof. Let $\sigma^k \in R_*(S_k)$ denote the homomorphism $R(S_k) \to \mathbf{Z}$ defined by $\qquad \sigma^k(1) = 1, \quad \sigma^k(V_\pi) = 0 \quad \text{if } V_\pi \neq 1,$

where 1 denotes the trivial representation. Since $\pi(E)$ is the kth symmetric power of E when $V_\pi = 1$, it follows from the definition of $\Delta'_{n,k}$ that

$$\Delta'_{n,k}(\sigma^k) = h_k(t_1,\ldots,t_n)$$

is the kth homogeneous symmetric function (i.e. the coefficient of z^k in $\prod (1-zt_i)^{-1}$). Since the h_k are a polynomial basis for the symmetric functions, it follows that Δ'_n is an epimorphism for all n. Now the rank of $R(S_k)$ is equal to the number of conjugacy classes of S_k, that is the number of partitions of k, and hence is also equal to the rank of $\mathrm{Sym}_k[t_1,\ldots,t_n]$ provided that $n \geqslant k$. Hence

$$\Delta'_{n,k}: R_*(S_k) \to \mathrm{Sym}_k[t_1,\ldots,t_n]$$

is an epimorphism of free abelian groups of the same rank (for $n \geqslant k$) and hence is an isomorphism. Since

$$\mathrm{Sym}_k[t_1,\ldots,t_{n+1}] \to \mathrm{Sym}_k[t_1,\ldots,t_n]$$

is also an isomorphism for $n \geqslant k$, this completes the proof.

Corollary 1.3. *R_* is a polynomial ring on generators $\sigma^1, \sigma^2,\ldots$.*

Instead of using the elements $\sigma^k \in R_*(S_k)$ we could equally well have used the elements λ^k defined by

$$\lambda^k(V_\pi) = 1 \quad \text{if } V_\pi \text{ is the sign representation.}$$
$$\lambda^k(V_\pi) = 0 \quad \text{otherwise.}$$

Since $\pi(E)$ is the kth exterior power when π is the sign representation of S_k, it follows that

$$\Delta'_{n,k}(\lambda^k) = e_k(t_1,\ldots,t_n)$$

is the kth elementary symmetric function. Thus R_* is equally well a polynomial ring on generators $\lambda^1, \lambda^2,\ldots$.

Corollary 1.4. *Let $\Delta_{n,k} = \sum a_i \otimes b_i$ with $a_i \in \mathrm{Sym}_k[t_1,\ldots,t_n]$ and $b_i \in R(S_k)$, and suppose $n \geqslant k$. Then the a_i form a base if and only if the b_i form a base. When this is so the a_i determine the b_i and conversely, i.e. they are 'dual bases'.*

Proof. This is an immediate reinterpretation of the fact that $\Delta'_{n,k}$ is an isomorphism.

Corollary 1.5. *The representations ρ_α form a base for $R(S_k)$.*

Proof. Apply Corollary 1.4 to the expression for $\Delta_{n,k}$ given in Proposition 1.1. Since the m_α are a basis for the symmetric functions, it follows that the ρ_α are a basis for $R(S_k)$.

Corollary 1.6. *The characters of S_k take integer values on all conjugacy classes.*

Proof. The characters of all ρ_α are integer-valued and so Corollary 1.6 follows from Corollary 1.5.

Note. Corollary 1.6 can of course be deduced fairly easily from other considerations.

Let $C(S_k)$ denote the group of integer-valued class functions on S_k. By Corollary 1.6 we have a natural homomorphism

$$R(S_k) \to C(S_k).$$

This has zero kernel and finite cokernel, and the same is therefore true for the dual homomorphism

$$C_*(S_k) \to R_*(S_k).$$

The direct sum $C_* = \sum_{k \geqslant 0} C_*(S_k)$ has a natural ring structure, and

$$C_* \to R_*$$

is a ring homomorphism. We shall identify C_* with the image subring of R_*. From its definition, $C_*(S_k)$ is the free abelian group on the conjugacy classes of S_k. Let ψ^k denote the class of a k-cycle. Then C_* is a polynomial ring on $\psi^1, \psi^2, \ldots$. The next result identifies the subring $\Delta'(C_*)$ of symmetric functions:

PROPOSITION 1.7. $\Delta'_n(\psi^k) = m_k(t_1, \ldots, t_n) = \sum_{i=1}^{n} t_i^k$ *so that* $\Delta'(C_*)$ *is the subring generated by the power sums* m_k.

Proof. By definition we have

$$\Delta'_n(\psi^k) = \mathrm{Trace}(gT^{\otimes k}),$$

where $g \in S_k$ is a k-cycle. Now use Proposition 1.1 to evaluate this trace and we get

$$\Delta'_n \psi^k = \sum_{\alpha \vdash k} m_\alpha \rho_\alpha(g).$$

But, if $H \subset G$, any character of G induced from H is zero on all elements of G not conjugate to elements of H. Hence, taking $H = S_\alpha$, $G = S_k$, we see that $\rho_\alpha(g) = 0$ unless $\alpha = k$ (i.e. α is the single partition k). Since $\rho_k(g) = 1$, we deduce

$$\Delta'_n \psi^k = m_k,$$

as required.

COROLLARY 1.8. *Let* Q_k *be the Newton polynomial expressing the power sum* m_k *in terms of the elementary symmetric functions* $e_1, \ldots, e_k$, *i.e.*

$$m_k = Q_k(e_1, \ldots, e_k),$$

then

$$\psi_k = Q_k(\lambda^1, \ldots, \lambda^k) \in R_*.$$

Remark. Let us tensor with the rationals $\mathbf{Q}$, so that we can introduce

$$\epsilon_\alpha \in R(S_k) \otimes \mathbf{Q},$$

the characteristic function of the conjugacy class defined by the partition α. Then Proposition 1.7 is essentially equivalent to the following expression [cf. (**11**) VII (7.6)] for $\Delta_{n,k}$

$$\Delta_{n,k} = \sum_{\alpha \vdash k} p_\alpha(t) \otimes \epsilon_\alpha \in \mathrm{Sym}_k[t_1,..., t_n] \otimes R(S_k) \otimes Q,$$

where p_α is the monomial in the power sums

$$p_\alpha = \prod_{i=1}^{k} (m_i)^{a_i}, \quad \alpha = 1^{a_1} 2^{a_2} ....$$

Since $\Delta'(\lambda_k) = e_k$, it follows that we can write $\Delta_{n,k}$ in the form

$$\Delta_{n,k} = \sum_{\alpha \vdash k} q_\alpha(t) \otimes b_\alpha,$$

where q_α is the monomial in the elementary symmetric functions

$$q_\alpha = \prod_{i=1}^{k} (e_i)^{a_i}, \quad \alpha = 1^{a_1} 2^{a_2} ...,$$

and the b_α are certain uniquely defined elements in $R(S_k)$. We shall not attempt to find b_α in general, but the following proposition gives the 'leading coefficient' b_k.

PROPOSITION 1.9. *Let M denote the $(k-1)$-dimensional representation of S_k given by the subspace $\sum_{i=1}^{k} z_i = 0$ of the standard k-dimensional representation. Let $\Lambda^i(M)$ denote the ith exterior power of M, and put*

$$\Lambda_{-1}(M) = \sum (-1)^i \Lambda^i(M) \in R(S_k).$$

Then we have

$$\Delta_{n,k} = (-1)^{k-1} e_k(t) \otimes \Lambda_{-1}(M) + \text{composite terms},$$

where 'composite' means involving a product of at least two $e_i(t)$.

Proof. In the formula

$$\Delta_{n,k} = \sum_{\alpha \vdash k} q_\alpha(t) \otimes b_\alpha,$$

the b_α are the basis of $R(S_k)$ dual to the basis of $R_*(S_k)$ consisting of monomials in the λ^i. Thus b_k is defined by the conditions

$$\langle b_k, \lambda^k \rangle = 1,$$
$$\langle b_k, u \rangle = 0$$

if u is composite in the λ^i. Since the ψ^i are related to the λ^i by the equations of Corollary 1.8

$$\psi^k = Q_k(\lambda^1,..., \lambda^k) = (-1)^{k-1} k \lambda_k + \text{composite terms},$$

we can equally well define b_k by the conditions

$$\langle b_k, \psi^k \rangle = (-1)^{k-1}k,$$
$$\langle b_k, u \rangle = 0$$

if u is composite in the ψ^i. To prove that $b_k = (-1)^{k-1}\Lambda_{-1}(M)$, it remains therefore to check that the character $\Lambda_{-1}(M)$ vanishes on all composite classes and has value k on a k-cycle. Now, if $g \in S_k$ is composite, i.e. not a k-cycle, it has an eigenvalue 1 when acting on M; if $g = (1 \ldots r)(r+1, \ldots s)\ldots$ is the cycle decomposition, the fixed vector is given by

$$z_i = \frac{1}{r} \quad (1 \leqslant i \leqslant r), \qquad z_j = -\frac{1}{k-r} \quad (j > r).$$

Since $\Lambda_{-1}(M)(g) = \det(1-g_M)$, where g_M is the linear transformation of M defined by g, the existence of an eigenvalue 1 of g_M implies $\Lambda_{-1}(M)(g) = 0$. Finally take $g = (1\ 2 \ldots k)$ and consider the k-dimensional representation $N = M \oplus 1$. Then g_N is given by the following matrix

$$g_N = \begin{pmatrix} 0 & 1 & & & \\ & & 1 & & \\ & & & \cdot & \\ & & & & \cdot \\ & & & & \cdot \quad 1 \\ 1 & & & & \end{pmatrix}$$

and so $\det(1-tg_N) = 1-t^k$. Hence

$$\det(1-tg_M) = \det(1-tg_N).(1-t)^{-1}$$
$$= \frac{1-t^k}{1-t} = 1+t+t^2+\ldots+t^{k-1},$$

and so $$\Lambda_{-1}(M)(g) = \det(1-g_M) = k,$$

which completes the proof.

If $G \subset S_k$ is any subgroup, then we can consider the element

$$\Delta_{n,k}(G) \in \mathrm{Sym}_k[t_1, \ldots, t_n] \otimes R(G)$$

obtained from $\Delta_{n,k}$ by the restriction $\eta: R(S_k) \to R(G)$. Similarly

$$\Delta'_{n,k}(G): R_*(G) \to \mathrm{Sym}_k[t_1, \ldots, t_n]$$

is the composition of $\Delta'_{n,k}$ and

$$\eta_*: R_*(G) \to R_*(S_k).$$

Consider in particular the special case when $k = p$ is *prime* and $G = Z_p$ is the cyclic group of order p. The image of

$$\eta: R(S_p) \to R(Z_p)$$

is generated by the trivial representation 1 and the regular representation N of Z_p (this latter being the restriction of the standard p-dimensional representation of S_p). Hence we must have

$$\Delta_{n,p}(Z_p) = a(t) \otimes 1 + b(t) \otimes N$$

for suitable symmetric functions $a(t)$, $b(t)$. Evaluating $R(S_p)$ on the identity element we get

$$e_1^p = a + pb.$$

Evaluating on a generator of Z_p and using Proposition 1.7 we get

$$m_p = a.$$

Hence $b = \frac{e_1^p - m_p}{p}$ which has, of course, integer coefficients since

$$(\sum t_i)^p \equiv \sum t_i^p \mod p.$$

Thus we have established the proposition:

Proposition 1.10. *Let p be a prime. Then restricting $\Delta_{n,p}$ from the symmetric group to the cyclic group we get*

$$\Delta_{n,p}(Z_p) = m_p \otimes 1 + \frac{e_1^p - m_p}{p} \otimes N,$$

where N is the regular representation of Z_p.

Let $\theta^p \in R_*(S_p)$ be the element corresponding to

$$\frac{e_1^p - m_p}{p} \in \mathrm{Sym}_p[t_1, \ldots, t_n]$$

by the isomorphism of Proposition 1.2 (for $n \geqslant p$), i.e.

$$\Delta_n' \theta^p = \frac{e_1^p - m_p}{p}.$$

Then Proposition 1.10 asserts that θ^p is that homomorphism $R(S_p) \to \mathbf{Z}$ which gives the multiplicity of the regular representation N when we restrict to Z_p. Thus, for $\rho \in R(S_p)$,

$$\eta(\rho) = \psi^p(\rho)1 + \theta^p(\rho)N, \tag{1.11}$$

where $\eta: R(S_p) \to R(Z_p)$ is the restriction.

2. Operations in K-theory

Let X be a compact Hausdorff space and let G be a finite group. We shall say that X is a *G-space* if G acts on X. Let E be a complex vector bundle over X. We shall say that E is a *G-vector bundle* over the G-space X if E is a G-space such that

(i) the projection $E \to X$ commutes with the action of G,

(ii) for each $g \in G$ the map $E_x \to E_{g(x)}$ is linear.

The Grothendieck group of all G-vector bundles over the G-space X is denoted by $K_G(X)$. Note that the action of G on X is supposed given: it is part of the structure of X. Since we can always construct an invariant metric in a G-vector bundle by averaging over G, the usual arguments show that a short exact sequence splits compatibly with G. Hence, if

$$0 \to E_1 \to E_2 \to \dots \to E_n \to 0$$

is a long exact sequence of G-vector bundles, the Euler characteristic $\sum(-1)^i[E_i]$ is zero in $K_G(X)$. For a fuller treatment of these and other points about $K_G(X)$ we refer the reader to (**4**) and (**9**).

In this section we shall be concerned only with a trivial G-space X, i.e. $g(x) = x$ for all $x \in X$ and $g \in G$. In this case a G-vector bundle is just a vector bundle E over X with a given homomorphism

$$G \to \operatorname{Aut} E,$$

where $\operatorname{Aut} E$ is the group of vector bundle automorphisms of E. We proceed to examine such a G-vector bundle.

The subspace of E left fixed by G forms a subvector bundle E^G of E: in fact it is the image of the projection operator

$$\frac{1}{|G|}\sum_{g\in G} g,$$

and the image of any projection operator is always a sub-bundle (**4**). If E, F are two G-vector bundles, then the subspace of $\operatorname{Hom}(E,F)$ consisting of all $\phi_x: E_x \to F_x$ commuting with the action of G forms a subvector bundle $\operatorname{Hom}_G(E,F)$: in fact $\operatorname{Hom}_G(E,F) = (\operatorname{Hom}(E,F))^G$. In particular let V be a representation space of G, and let $\mathbf{V}$ denote the corresponding G-vector bundle $X \times V$ over X. Then, for any G-vector bundle E over X, $\operatorname{Hom}_G(\mathbf{V}, E)$ is a vector bundle, and we have a natural homomorphism $\mathbf{V} \otimes \operatorname{Hom}_G(\mathbf{V}, E) \to E$.

Now let $\{V_\pi\}$... be a complete set of irreducible representations of G and consider the bundle homomorphism

$$\alpha: \sum_\pi \{\mathbf{V}_\pi \otimes \operatorname{Hom}_G(\mathbf{V}_\pi, E)\} \to E.$$

For each $x \in X$, α_x is an isomorphism. Hence α is an isomorphism. This establishes the following proposition:

Proposition 2.1. *If X is a trivial G-space, we have a natural isomorphism*

$$K(X) \otimes R(G) \to K_G(X).$$

In particular we can apply the preceding discussion to the natural action of S_k on the k-fold tensor product $E^{\otimes k}$ of a vector bundle E. Thus we have a canonical decomposition compatible with the action of S_k

$$E^{\otimes k} \cong \sum_{\pi} \{V_\pi \otimes \mathrm{Hom}_{S_k}(V_\pi, E^{\otimes k})\}.$$

We put
$$\pi(E) = \mathrm{Hom}_{S_k}(V_\pi, E^{\otimes k}).$$

Thus π is an operation on vector bundles. In fact $\pi(E)$ is the vector bundle associated to E by the irreducible representation of $GL(n)$ ($n = \dim E$) associated to the partition π, but this fact will play no special role in what follows.

Our next step is to extend these operations on vector bundles to operations on $K(X)$. For this purpose it will be convenient to represent $K(X)$ as the quotient of a set $\mathscr{C}(X)$ by an equivalence relation (elements of $\mathscr{C}(X)$ will play the role of 'cochains'). An element of $\mathscr{C}(X)$ is a *graded* vector bundle $E = \sum_{i \in Z} E_i$, where $E_i = 0$ for all but a finite number of values of i. We have a natural surjection

$$\mathscr{C}(X) \to K(X)$$

given by taking the Euler characteristic $[E] = \sum (-1)^i [E_i]$. The equivalence relation on $\mathscr{C}(X)$ which gives $K(X)$ is clearly generated by isomorphism and the addition of *elementary* objects, i.e. one of the form $\sum P_i$ with

$$P_j = P_{j+1} \quad (\text{for some } j), \qquad P_i = 0 \quad (i \neq j, j+1).$$

Similarly for a G-space X we can represent $K_G(X)$ as a quotient of $\mathscr{C}_G(X)$, where an element of $\mathscr{C}_G(X)$ is a graded G-vector bundle.

Suppose now that $E \in \mathscr{C}(X)$ is a graded vector bundle. Then $E^{\otimes k}$ is also a graded vector bundle, the grading being defined in the usual way as the sum of the degrees of the k factors. We consider S_k as acting on $E^{\otimes k}$ by permuting factors and with the *appropriate sign change*. Thus a transposition of two terms $e_p \otimes e_q$ (where $e_p \in E_p$, $e_q \in E_q$) carries with it the sign $(-1)^{pq}$. The Euler characteristic $[E^{\otimes k}]$ of $E^{\otimes k}$ is then an element of $K_{S_k}(X)$.

PROPOSITION 2.2. *The element* $[E^{\otimes k}] \in K_{S_k}(X)$ *depends only on the element* $[E] \in K(X)$. *Thus we have an operation*:

$$\otimes k : K(X) \to K_{S_k}(X) = K(X) \otimes R(S_k).$$

Proof. We have to show that, if P is an elementary object of $\mathscr{C}(X)$, then

$$[(E \oplus P)^{\otimes k}] = [E^{\otimes k}] \in K_{S_k}(X).$$

But we have an S_k-decomposition:

$$(E \oplus P)^{\otimes k} \cong E^{\otimes k} \oplus Q.$$

We have to show therefore that $[Q] = 0$ in $K_{S_k}(X)$. To do this we regard E as a *complex* of vector bundles with all maps zero and P as a complex with the identity map $P_j \to P_{j+1}$. Then $(E \oplus P)^{\otimes k}$ is a complex of vector bundles, and S_k acts on it as a group of complex automorphisms (because of our choice of signs). The same is true for $E^{\otimes k}$ and Q. Now Q contains P as a factor, and so Q is certainly acyclic. Hence, by the remark at the beginning of this section, we have $[Q] = 0$ in $K_{S_k}(X)$ as required.

Remark. If we decompose $E^{\otimes k}$ under S_k
$$E^{\otimes k} \cong \sum_{\pi} \mathbf{V}_\pi \otimes \pi(E),$$
where $\pi(E) = \mathrm{Hom}_{S_k}(\mathbf{V}_\pi, E^{\otimes k})$, Proposition 2.2 asserts that $E \mapsto \pi(E)$ induces an operation
$$\pi: K(X) \to K(X).$$

Let $\mathrm{Op}(K)$ denote the set of all natural transformations of the functor K into itself. In other words, an element $T \in \mathrm{Op}(K)$ defines for each X a map
$$T(X): K(X) \to K(X),$$
which is natural. We define addition and multiplication in $\mathrm{Op}(K)$ by adding and multiplying values. Thus, for $a \in K(X)$,
$$(T+S)(X)(a) = T(X)(a) + S(X)a,$$
$$TS(X)(a) = T(X)a \,.\, S(X)a.$$

If we follow the operation
$$\otimes k: K(X) \to K(X) \otimes R(S_k)$$
by a homomorphism $\phi: R(S_k) \to \mathbf{Z}$ we obtain a natural map
$$T_\phi: K(X) \to K(X).$$
This procedure defines a map
$$j_k: R_*(S_k) \to \mathrm{Op}(K)$$
which is a group homomorphism. Extending this additively we obtain a *ring homomorphism*
$$j: R_* \to \mathrm{Op}(K).$$
We have now achieved our aim of showing how the symmetric group defines a ring of operations in K-theory. The structure of the ring R_* has moreover been completely determined in § 1. We conclude this section by examining certain particular operations and connecting up our definitions of them with those given by Grothendieck [cf (**5**); § 12] and Adams (**2**).

To avoid unwieldy formulae we shall usually omit the symbol j and just think of elements of R_* as operations. In fact it is not difficult to show that j is a monomorphism (although we do not really need this fact), so that R_* may be thought of as a subring of $\mathrm{Op}(K)$.

All the particular elements that we have described in § 1, namely $\sigma^k, \lambda^k, \psi^k, \theta^p$, can now be regarded as operations in K-theory. From the way they were defined it is clear that, if E is vector bundle, then $\lambda^k[E]$ is the class of the kth exterior power of E, and $\sigma^k(E)$ is the class of the kth symmetric power of E. A general element of $K(X)$ can always be represented in the form $[E_0]-[E_1]$, where E_0, E_1 are vector bundles. Taking $(E_0 \oplus E_1)^{\otimes k}$ as an S_k-complex and picking out the symmetric and skew-symmetric components, we find

$$\sigma^k([E_0]-[E_1]) = \sum_{j=0}^{k} (-1)^j \sigma^{k-j}[E_0]\lambda^j[E_1], \tag{1}$$

$$\lambda^k([E_0]-[E_1]) = \sum_{j=0}^{k} (-1)^j \lambda^{k-j}[E_0]\sigma^j[E_1]. \tag{2}$$

Putting formally $\lambda_u = \sum \lambda^k u^k$, $\sigma_u = \sum \sigma^k u^k$, where u is an indeterminate, and taking $E_0 = E_1$ in (1), we get

$$\sigma_u[E_1]\lambda_{-u}[E_1] = 1. \tag{3}$$

This identity could of course have been deduced from the corresponding relation between the generating functions of e_k and h_k by using the isomorphism of (1.2). Now from (2) we get

$$\begin{aligned}\lambda_u([E_0]-[E_1]) &= \lambda_u[E_0]\sigma_{-u}[E_1] \\ &= \lambda_u[E_0]\lambda_u[E_1]^{-1} \quad \text{by (3)}.\end{aligned}$$

This is the formula by which Grothendieck originally extended the λ^k from vector bundles to K. Thus our definition of the operations λ^k coincides with that of Grothendieck. Essentially the use of graded tensor products has provided us with a general procedure for extending operations which can be regarded as a generalization of the Grothendieck method for the exterior powers.†

Adams defines his operations ψ^k in terms of the Grothendieck λ^k by use of the Newton polynomials

$$\psi^k = Q_k(\lambda^1, \ldots, \lambda^k).$$

Corollary 1.8 shows that our definition of ψ^k therefore agrees with that of Adams. An important property of the ψ^k is that they are additive. We shall therefore show how to prove this directly from our definition.

PROPOSITION 2.3. *Let E, F be vector bundles, then*

$$\psi^k([E]\pm[F]) = \psi^k[E]\pm\psi^k[F].$$

† This fact was certainly known to Grothendieck.

Proof. Construct a graded vector bundle D with $D_0 = E, D_1 = F$ and consider $D^{\otimes k}$. The same reasoning as used in Proposition 1.1 shows that

$$[D]^{\otimes k} = \sum_{j=0}^{k} (-1)^j \operatorname{ind}_j[E^{\otimes k-j} \otimes F^{\otimes j}] \in K(X) \otimes R(S_k),$$

where $\operatorname{ind}_j : K(X) \otimes R(S_{k-j} \times S_j) \to K(X) \otimes R(S_k)$ is given by the induced representation. Here $E^{\otimes k-j}$ is an S_{k-j}-vector bundle via the standard permutation, while S_j acts on $F^{\otimes j}$ via permutation and signs. To obtain $\psi^k[D]$ we have to evaluate $R(S_k)$ on a k-cycle. As in Proposition 1.1 all terms except $j = 0, k$ give zero; since the sign of a k-cycle is $(-1)^{k-1}$ we get

$$\begin{aligned}\psi^k([E]-[F]) &= \psi^k[E]+(-1)^k(-1)^{k-1}\psi^k[F] \\ &= \psi^k[E]-\psi^k[F].\end{aligned}$$

For $[E]+[F]$ the argument is similar but easier.

The multiplicative property

$$\psi^k[E \otimes F] = \psi^k[E]\psi^k[F]$$

follows at once from the isomorphism

$$(E \otimes F)^{\otimes k} \cong E^{\otimes k} \otimes F^{\otimes k}$$

and the multiplicative property of the trace.

Suppose now that we have any expansion, as in Corollary 1.4, of the basic element $\Delta_{n,k}$ in the form

$$\Delta_{n,k} = \sum a_i \otimes b_i,$$

where the $a_i \in \mathrm{Sym}_k[t_1,..., t_n]$ are a basis and the $b_i \in R(S_k)$ are therefore a dual basis (assuming $n \geqslant k$). Then, for any $x \in K(X)$, we obtain a corresponding expansion for $x^{\otimes k}$:

$$x^{\otimes k} = \alpha_i(x) \otimes b_i \in K(X) \otimes R(S_k),$$

where $\alpha_i = (\Delta')^{-1}a_i \in R_*$. This follows at once from the definition of Δ' and the way we have made R_* operate on $K(X)$.

Taking the a_i to be the monomials in the elementary symmetric functions the α_i are then the corresponding monomials in the exterior powers λ^i. Proposition 1.9 therefore gives the following proposition:†

PROPOSITION 2.4. *For any $x \in K(X)$ we have*

$$x^{\otimes k} = (-1)^{k-1}\lambda^k(x) \otimes \lambda_{-1}(M) + \text{composite terms},$$

where 'composite' means involving a product of at least two $\lambda^i(x)$ and M is the $(k-1)$-dimensional representation of S_k.

† Now that we have identified the λ^i of § 1 with the exterior powers we revert to the usual notation and write $\lambda^i(M)$ instead of $\Lambda^i(M)$, and correspondingly $\lambda_{-1}(M)$ instead of $\Lambda_{-1}(M)$.

Now let us restrict ourselves to the cyclic group Z_k. The image of $x^{\otimes k}$ in $K(X) \otimes R(Z_k)$ will be denoted by $P^k(x)$ and called the *cyclic kth power.* In the particular case when $k = p$ (a prime), (1.11) leads to the following proposition:

PROPOSITION 2.5. *Let p be a prime and let $x \in K(X)$. Then the cyclic pth power $P^p(x)$ is given by the formula*

$$P^p(x) = \psi^p(x) \otimes 1 + \theta^p(x) \otimes N \in K(X) \otimes R(Z_p),$$

where N is the regular representation of Z_p.

Now ψ^p and θ^p correspond, under the isomorphism

$$\Delta' : R_* \to \lim_{\substack{\leftarrow \\ n}} \mathrm{Sym}[t_1, \ldots, t_n],$$

to the polynomials $\sum t_i^p$ and $\dfrac{(\sum t_i)^p - \sum t_i^p}{p}$ respectively. Hence they are related by the formula

$$\psi^p = (\psi^1)^p - p\theta^p,$$

so that, for any $x \in K(X)$, we have

$$\psi^p(x) = x^p - p\theta^p(x).$$

Substituting this in (2.5) we get the formula

$$P^p(x) = x^p \otimes 1 + \theta^p(x) \otimes (N - p). \tag{2.6}$$

This is a better way of writing (2.5) since it corresponds to the decomposition

$$R(Z_p) = \mathbf{Z} \oplus I(Z_p),$$

where $I(Z_p)$ is the augmentation ideal. Thus

$$\theta^p(x) \otimes (N - p) \in K(X) \otimes I(Z_p)$$

represents the difference between the pth cyclic power $P^p(x)$ and the 'ordinary' pth power $x^p \otimes 1$.

Proposition 2.5 leads to a simple geometrical description for $\psi^p[V]$, where V is a vector bundle. Let T be the automorphism of $V^{\otimes p}$ which permutes the factors cyclically and V_j be the eigenspace of T corresponding to the eigenvalue $\exp(2\pi i j/p)$. Then

$$\psi^p[V] = [V_0] - [V_1]. \tag{2.7}$$

In fact from Proposition 2.5 we see that

$$[V_0] = \psi^p[V] + \theta^p[V],$$

$$[V_j] = \theta^p[V] \quad (j = 1, \ldots, p-1).$$

3. External tensor powers

For a further study of the properties of the operation $\otimes k$ it is necessary both to 'relativize' it and to 'externalize' it.

First consider the relative group $K_G(X, Y)$, where X is a G-space, Y a sub G-space. As with the absolute case we can consider $K_G(X, Y)$ as the quotient of a set $\mathscr{C}_G(X, Y)$ by an equivalence relation. An object E of $\mathscr{C}_G(X, Y)$ is a G-complex of vector bundles over X acyclic over Y, i.e. E consists of G-vector bundles E_i (with $E_i = 0$ for all but a finite number) and homomorphisms

$$\to E_i \xrightarrow{d} E_{i+1} \xrightarrow{d}$$

commuting with the action of G, so that $d^2 = 0$ and over each point of Y the sequence is exact. An elementary object P is one in which $P_i = 0$ $(i \neq j, j+1)$, $P_j = P_{j+1}$, and $d: P_j \to P_{j+1}$ is the identity. The equivalence relation imposed on $\mathscr{C}_G(X, Y)$ is that generated by isomorphism and addition (direct sum) of elementary objects. Then, if $E \in \mathscr{C}_G(X, Y)$, its equivalence class $[E] \in K_G(X, Y)$. For the details we refer to (**4**). For the analogous results in the case when there is no group, i.e. for the definition of $K(X, Y)$ as a quotient of $\mathscr{C}(X, Y)$, we refer to (**7**) [Part II].

Consider next the *external* tensor power. If E is a vector bundle over X, we define $E^{\boxtimes k}$ to be the vector bundle over the Cartesian product X^k (k factors of X) whose fibre at the point $(x_1 \times x_2 \times \ldots \times x_k)$ is $E_{x_1} \otimes E_{x_2} \otimes \ldots \otimes E_{x_k}$. Thus $E^{\boxtimes k}$ is an S_k-vector bundle over the S_k-space X^k, the symmetric group S_k acting in the usual way on X^k by permuting the factors. Clearly, if

$$d: X \to X^k$$

is the diagonal map, we have a natural S_k-isomorphism

$$d^*(E^{\boxtimes k}) \cong E^{\otimes k}. \tag{3.1}$$

If E is a complex of vector bundles over X, then we can define in an obvious way $E^{\boxtimes k}$, which will be a complex of vector bundles over X^k. Moreover $E^{\boxtimes k}$ will be an S_k-complex of vector bundles, X^k being an S_k-space as above. If E is acyclic over $Y \subset X$, then $E^{\boxtimes k}$ will be acyclic over the subspace of X consisting of points $(x_1 \times x_2 \times \ldots \times x_k)$ with $x_i \in Y$ for at least one value of i. We denote this subspace by $X^{k-1}Y$ and we write $(X, Y)^k$ for the pair $(X^k, X^{k-1}Y)$. Thus we have defined an operation

$$\boxtimes k: \mathscr{C}(X, Y) \to \mathscr{C}_{S_k}(X, Y)^k.$$

The proof of (2.2) generalizes at once to this situation and establishes

PROPOSITION 3.2. *The operation* $E \mapsto E^{\boxtimes k}$ *induces an operation*

$$\boxtimes k : K(X, Y) \to K_{S_k}(X, Y)^k.$$

COROLLARY 3.3. *If x is in the kernel of $K(X) \to K(Y)$, then $x^{\boxtimes k}$ is in the kernel of*

$$K_{S_k}(X^k) \to K_{S_k}(X^{k-1}Y).$$

Proof. This follows at once from (3.2) and the naturality of the operation $\boxtimes k$.

From (3.1) we obtain the commutative diagram

$$\begin{array}{ccc} K(X) & \xrightarrow{\boxtimes k} & K_{S_k}(X^k) \\ & \searrow^{\otimes k} & \downarrow d^* \\ & & K_{S_k}(X) \end{array} \qquad (3.4)$$

4. Operations and filtrations

From now we assume that the spaces $X, Y, \ldots$ are *finite CW-complexes.* Then $K(X)$ is filtered by the subgroups $K_q(X)$ defined by

$$K_q(X) = \operatorname{Ker}\{K(X) \to K(X_{q-1})\},$$

where X_{q-1} denotes the $(q-1)$-skeleton of X. Thus $K_0(X) = K(X)$ and $K_n(X) = 0$ if $\dim X < n$. Moreover, as shown in (**8**), we have

$$K_{2q}(X) = K_{2q-1}(X)$$

for all q. Since any map $Y \to X$ is homotopic to a cellular map, it follows that the filtration is natural.

In [**8**] it is shown that $K(X)$ is a *filtered ring*, i.e. that $K_p K_q \subset K_{p+q}$. In particular it follows that

$$x \in K_q(X) \Rightarrow x^k \in K_{kq}(X).$$

We propose to generalize this result to the tensor power $\otimes k$.

We start by recalling (**5**) that, for any finite group, there is a natural homomorphism

$$\alpha : R(G) \to K(B_G),$$

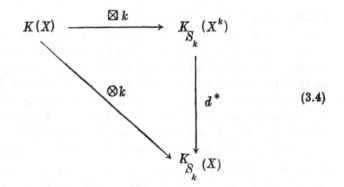

where B_G is the classifying space of G. This homomorphism arises as follows. Let A be the universal covering of B_G and V be any G-module. Then $A \times_G V$ is a vector bundle over B_G. The construction $V \mapsto A \times_G V$ induces the homomorphism

$$\alpha: R(G) \to K(B_G).$$

This construction can be generalized as follows. Let X be a G-space and denote by X_G the space $A \times_G X$. If V is a G-vector bundle over X, then

$$V_G = A \times_G V$$

is a vector bundle over X_G. The construction $V \mapsto V_G$ then induces a homomorphism

$$\alpha_X: K_G(X) \to K(X_G).$$

A couple of remarks are needed here. In the first place there is a clash of notation concerning B_G. To fit in with our general notation we should agree that 'B' is a point space. Secondly X_G, like B_G, is not a finite complex. Now B_G can be taken as an infinite complex in which the q-skeleton $B_{G,q}$ is finite for each q, and $K(B_G)$ can be defined by

$$K(B_G) = \lim_{\overleftarrow{q}} K(B_{G,q}).$$

If we suppose that G acts cellularly on X, then we can put $X_{G,q} = A_q \times_G X$, where A_q is the universal covering of $B_{G,q}$ and $X_{G,q}$ will be a finite complex. We then define

$$K(X_G) = \lim_{\leftarrow} K(X_{G,q}).$$

In fact, as will become apparent, there is no need for us to proceed to the limit. All our results will essentially be concerned with finite skeletons. We have introduced the infinite spaces B_G, X_G because it is a little tidier than always dealing with finite approximations.

Applying the above to the group S_k and the spaces X (trivial action) and X^k (permutation action) we obtain a commutative diagram

$$\begin{array}{ccc}
K_{S_k}(X^k) & \xrightarrow{\alpha_{X^k}} & K(X^k_{S_k}) \\
\downarrow{\scriptstyle d^*} & & \downarrow{\scriptstyle d^*} \\
K_{S_k}(X) & \xrightarrow{\alpha_X} & K(X_{S_k}) \\
\| & & \| \\
K(X) \otimes R(S_k) & \longrightarrow & K(X \times B_{S_k}),
\end{array} \qquad (4.1)$$

where d^* is induced by the diagonal map $d: X \to X^k$.

Proposition 4.2. *Let $x \in K_q(X)$, then*

$$\alpha_{X^k}(x^{\boxtimes k}) \in K_{kq}(X^k_{S_k}).$$

Proof. By hypothesis x is in the kernel of

$$K(X) \to K(X_{q-1}).$$

Hence applying (3.3) with $Y = X_{q-1}$ we deduce that $x^{\boxtimes k}$ is in the kernel of ρ in the following diagram

$$\begin{array}{ccc} K_{S_k}(X^k) & \xrightarrow{\alpha_{X^k}} & K(X^k_{S_k}) \\ \big\downarrow{\scriptstyle \rho} & & \big\downarrow \\ K_{S_k}(X^{k-1}X_{q-1}) & \longrightarrow & K((X^{k-1}X_{q-1})_{S_k}) \end{array}$$

The required result now follows from this diagram, provided that we verify that

$$(X^k_{S_k})_{kq-1} \subset (X^{k-1}X_{q-1})_{S_k}.$$

But any cell σ of the $(kq-1)$-skeleton of $X^k_{S_k} = X^k \times_{S_k} A$ arises from a product of k cells of X and a cell of A. Hence at least one of the cells of X occurring must have dimension less than q, and so σ is contained in

$$(X^{k-1}X_{q-1})_{S_k} = X^{k-1}X_{q-1} \times_{S_k} A,$$

as required.

Since the filtration in K is natural, Proposition 4.2 together with the diagram (4.1) and Corollary 3.3 gives our main result:

THEOREM 4.3. *Let $\otimes k: K(X) \to K(X) \otimes R(S_k)$ be the tensor power operation, and let*

$$\alpha: K(X) \otimes R(S_k) \to K(X \times B_{S_k})$$

be the natural homomorphism. Then

$$x \in K_q(X) \Rightarrow \alpha(x^{\otimes k}) \in K_{kq}(X \times B_{S_k}).$$

COROLLARY 4.4. *Let $\dim X \leqslant n$ and let $x \in K_q(X)$. Then the image of $x^{\otimes k}$ in $K(X) \otimes K(B_{S_k, kq-n-1})$ is zero.*

Proof. By Theorem 4.3 $x^{\otimes k}$ has zero image in $K(X \times B_{S_k, kq-n-1})$. But for any two spaces A, B the map

$$K(A) \otimes K(B) \to K(A \times B)$$

is injective (6). Hence $x^{\otimes k}$ gives zero in $K(X) \otimes K(B_{S_k, kq-n-1})$ as required.

Remark. Theorem 4.3 suggests that for any finite group G and G-space X we should define a filtration on $K_G(X)$ by putting

$$K_G(X)_q = \alpha_X^{-1} K_q(X \times B_G).$$

With this notation Theorem 4.3 would read simply

$$x \in K_q(X) \Rightarrow x^{\otimes k} \in K_{S_k}(X)_{kq}.$$

To exploit Theorem 4.3 we really need to know the filtration on $K(B_{S_k})$ as is shown by the following theorem:

Theorem 4.5. *Assume that $K(X)$ is torsion-free and let* $\dim X \leqslant n$. *Let* $x \in K_q(X)$ *and assume that all products* $\lambda^i(x)\lambda^j(x)$ *with* $i, j > 0$, $i+j \leqslant k$ *vanish. Then* $\lambda^k(x)$ *is divisible by the least integer* m *for which*

$$m\alpha\lambda_{-1}(M) \in K_{kq-n}(B_{S_k}),$$

M being as in Proposition 2.4. *In particular this holds in the stable range* $n < 2q$.

Proof. The hypotheses and Proposition 2.4 imply that

$$x^{\otimes k} = (-1)^{k-1}\lambda^k(x) \otimes \lambda_{-1}(M) \in K(X) \otimes R(S_k).$$

Let $A = K(B_{S_k})/K_{kq-n}(B_{S_k})$, so that A is a subgroup of $K(B_{S_k,kq-n-1})$. From Corollary 4.4 and the fact that $K(X)$ is free it follows that the image of $x^{\otimes k}$ in $K(X) \otimes A$ must be zero. Hence $\lambda^k(x)$ must be divisible by the order of the image of $\lambda_{-1}(M)$ in A, i.e. by the least integer m for which

$$m\alpha\lambda_{-1}(M) \in K_{kq-n}(B_{S_k}).$$

Remark. In the proof of Proposition 1.9 we saw that the character of $\lambda_{-1}(M)$ vanishes on all composite cycles of S_k. Thus, if k is not a prime-power, the character of $\lambda_{-1}(M)$ vanishes on all elements of S_k of prime-power order and so by (5) [(6.10)] $\lambda_{-1}(M)$ is in the kernel of the homomorphism

$$R(S_k) \to \widehat{R(S_k)}.$$

Hence $\alpha\lambda_{-1}(M) = 0$ and so Theorem 4.5 becomes vacuous. *Thus Theorem 4.5 is of interest only when k is a prime-power.*

In order to obtain explicit results it is necessary to restrict from S_k to the cyclic group Z_k. In this case the calculations are simple. First we need the lemma:

Lemma 4.6. *Let* $Y = B_{Z_k}$, *then*

$$K(Y_{2q-1}) \cong R(Z_k)/I(Z_k)^q.$$

Proof. Since Y has no odd integer cohomology, it follows that $K^1(Y, Y_{2q-1}) = 0$, and so from the exact sequence of this pair we deduce

$$K(Y_{2q-1}) \cong K(Y)/K_{2q}(Y).$$

But we know [(5) (8.1)] that

$$K(Y) \cong \widehat{R(Z_k)},$$

and $K_{2q}(Y)$ is the ideal generated by $I(Z_k)^q$. Hence

$$K(Y)/K_{2q}(Y) \cong R(Z_k)/I(Z_k)^q,$$

and the lemma is established.

Remark. The results quoted from (5) are quite simple, and we could easily have applied the calculations used there directly to Y_{2q-1}.

Combining Corollary 4.4 and Lemma 4.6 we deduce the proposition:

PROPOSITION 4.7. *Let* $\dim X \leqslant 2m$ *and let* $x \in K_{2q}(X)$. *Then the kth cyclic power* $P^k(x) \in K(X) \otimes R(Z_k)$ *is in the image of* $K(X) \otimes I(Z_k)^{kq-m}$.

The case when $k = p$, a prime, is of particular interest because Z_p is then the p-Sylow subgroup of S_p. This means that, as far as p-primary results go, nothing is lost on passing from S_p to Z_p. In the next section therefore we shall study this case in detail.

5. The prime cyclic case

LEMMA 5.1. *Let* $\rho \in R(Z_p)$ *denote the canonical one-dimensional representation of* Z_p,

$$N = \sum_{i=0}^{p-1} \rho^i$$

the regular representation and $\eta = \rho - 1$.

Then in $\widehat{R(Z_p)}$ *we have*

$$p^k(N-p) = (-1)^k \eta^{(k+1)(p-1)} + \text{higher terms}.$$

Proof. Since $\rho^p = 1$, we have $(1+\eta)^p = 1$. Thus $\eta^p = -p\eta\epsilon$, where $\epsilon \equiv 1 \bmod \eta$ and so is a unit in $\hat{R}$. Hence

$$(-p)\eta \sim \eta^p, \tag{1}$$

where we write $a \sim b$ if $a = \epsilon b$ with $\epsilon \equiv 1 \bmod \eta$. Now the identity

$$\sum_{i=0}^{p-1} (1+t)^i = \frac{(1+t)^p - 1}{t} \equiv p + t^{p-1} \quad \bmod pt$$

with t replaced by η shows that

$$\begin{aligned} N-p &\equiv \eta^{p-1} \quad \bmod p\eta \\ &\equiv \eta^{p-1} \quad \bmod \eta^p \quad \text{by (1)}. \end{aligned}$$

Hence we have $$(N-p) \sim \eta^{p-1}. \tag{2}$$

From (1) we have $$(-p)^k \eta \sim \eta^{k(p-1)}\eta,$$

and so $$(-p)^k \eta^{p-1} \sim \eta^{(k+1)(p-1)}. \tag{3}$$

The lemma now follows from (2) and (3).

COROLLARY 5.2. *The order of the image of* $(N-p)$ *in* $R(Z_p)/I(Z_p)^n$ *is* p^k *where k is the least integer such that* $k+1 \geqslant \dfrac{n}{p-1}$.

Proof. $I(Z_p)$ is the ideal (η).

We can now state the explicit result for the prime case:

THEOREM 5.3. *Suppose that* $\dim X \leqslant 2(q+t)$ *with* $t < q(p-1)$ *and let* $x \in K_{2q}(X)$. *Then* $\theta^p(x)$ *is divisible by* p^{q-r-1}, *where*

$$r = \left[\frac{t}{p-1}\right].$$

Proof. Since $\dim X < 2qp$, we have $x^p = 0$. Hence by Proposition 2.5 we have

$$P^p(x) = \theta^p(x) \otimes (N-p) \in K(X) \otimes R(Z_p).$$

By Proposition 4.7 it follows that $\theta^p(x)$ is divisible by the order of the image of $(N-p)$ in $R(Z_p)/I(Z_p)^n$, where

$$n = pq-q-t.$$

From Theorem 5.3 it follows that $\theta^p(x)$ is divisible by p^k, where k is the least integer for which

$$(k+1) \geqslant q - \frac{t}{p-1},$$

namely
$$k = q - \left[\frac{t}{p-1}\right] - 1.$$

COROLLARY 5.4. *Let the hypotheses be the same as in Theorem* 5.3. *Then* $\psi^p(x)$ *is divisible by* p^{q-r}, *where* $r = \left[\frac{t}{p-1}\right]$.

Proof. ψ^p and θ^p are related by the formula

$$\psi^p(x) = x^p - p\theta^p(x).$$

Since $x^p = 0$ in our case, we have

$$\psi^p(x) = -p\theta^p(x),$$

and so the result follows at once from Corollary 5.2.

Remark. Taking $t = 0$ we find that $\psi^p(x)$ is divisible by p^q on the sphere S^{2q}. Note that this result was not fed in explicitly anywhere. It is of course a consequence of the periodicity theorem, and the computation we have used for $K(B_{Z_p})$ naturally depended on the periodicity theorem.

The preceding results take a rather interesting form if X has no torsion. First we need a lemma:

LEMMA 5.5. *Suppose that X has no torsion (i.e. $H^*(X, \mathbf{Z})$ has no torsion) and let $x \in K(X)$. Suppose that the image of x in $K(X_q)$ is divisible by* d. *Then x is divisible by d* modulo $K_{q+1}(X)$, *i.e.*

$$x = dy+z, \quad y \in K(X), \quad z \in K_{q+1}(X).$$

Proof. Let A, B denote the image and cokernel of

$$j^*: K(X) \to K(X_q).$$

From the exact sequence of the pair (X, X_q) we see that B is isomorphic to a subgroup of $K^1(X, X_q)$. But, since X is torsion-free, so is X/X_q. Hence $K^1(X, X_q)$ is free and therefore also B. Hence, if $a \in A$ is divisible by d in $K(X_q)$, it is also divisible by d in A. Taking $a = j^*(x)$ therefore we have

$$j^*(x) = dj^*(y) \quad \text{for some } y \in K(X),$$

and so $x = dy+z$, for some $z \in \operatorname{Ker} j^* = K_{q+1}(X)$.

Using this lemma we now show how Corollary 5.4 leads to the following proposition:

PROPOSITION 5.6. *Suppose that X has no torsion and let $x \in K_{2q}(X)$. Then there exist elements*

$$x_i \in K_{2q+2i(p-1)}(X) \quad (i = 0, 1, \ldots, q)$$

such that

$$\psi^p(x) = \sum_{i=0}^{q} p^{q-i} x_i,$$

Moreover we can choose $x_q = x^p$.

Proof. By Theorem 5.3 the restriction of $\psi^p(x)$ to the $2(q+t)$-skeleton, with $t = i(p-1)-1$, is divisible by p^{q-i+1}. By Corollary 5.4 it follows that $\psi^p(x)$ is divisible by p^{q-i+1} modulo $K_{2q+2i(p-1)}(X)$. The required result now follows by induction on i. Since $\psi^p(x) \equiv x^p \bmod p$ and $x^p \in K_{2pq}(X)$, it follows that x^p is a choice for x_q.

The elements x_i occurring in Lemma 5.6 are not uniquely defined by x. If, however, we pass to the associated graded group $GK^*(X)$ and then reduce mod p, we see that the element

$$\bar{x}_i \in G^{2q+2i(p-1)}K(X) \otimes Z_p$$

defined by x_i *is* uniquely determined from the relation

$$\psi^p x = \sum_{i=0}^{q} p^{q-i} x_i.$$

If we multiply x by p or add to it anything in $K_{2(q+1)}(X)$, we see from Lemma 5.5 that $\bar{x}_i$ is unchanged. Hence $\bar{x}_i$ depends only on

$$\bar{x} \in G^{2q}K(X) \otimes Z_p.$$

Now we recall [(8) § 2] that, since X has no torsion, we have an isomorphism of graded rings

$$H^*(X, \mathbf{Z}) \cong GK^*(X),$$

and hence

$$H^{2q}(X, Z_p) \cong G^{2q}K(X) \otimes Z_p.$$

By this isomorphism the operation $\bar{x} \to \bar{x}_i$ must correspond to some cohomology operation. In the next section we shall show that this is precisely the Steenrod power P_p^i.

6. Relation with cohomology operations

In the proof of Proposition 4.2 we verified that there was an inclusion

$$j:(X^k, X^k_{2kq-1}) \to (X, X_{2q-1})^k.$$

Hence we can consider the map

$$K(X, X_{2q-1}) \to K(X^k_{S_k}, (X^k_{S_k})_{2kq-1})$$

given by $x \longmapsto \alpha j^* x^{\boxtimes k}$. If we follow this by a cellular approximation to the diagonal map $X_{S_k} \to X^k_{S_k}$, we obtain a map

$$\mu: K(X, X_{2q-1}) \to K(X_{S_k}, (X_{S_k})_{2kq-1}).$$

From its definition this is compatible with the operation

$$x \longmapsto d^* \alpha x^{\boxtimes k} = \alpha x^{\otimes k}$$

for the absolute groups, i.e. we have a commutative diagram

$$\begin{array}{ccc} K(X, X_{2q-1}) & \longrightarrow & K(X_{S_k}, (X_{S_k})_{2kq-1}) \\ \downarrow & & \downarrow \\ K(X) & \longrightarrow & K(X_{S_k}) \end{array} \tag{6.1}$$

On the other hand, by restricting X to X_{2q} and X_{S_k} to $(X_{S_k})_{2kq}$ we obtain another commutative diagram

$$\begin{array}{ccc} K(X, X_{2q-1}) & \xrightarrow{\mu} & K(X_{S_k}, (X_{S_k})_{2kq-1}) \\ \downarrow & & \downarrow \\ K(X_{2q}, X_{2q-1}) & & K((X_{S_k})_{2kq}, (X_{S_k})_{2kq-1}) \\ \| & & \| \\ C^{2q}(X) & \xrightarrow{\nu} & C^{2kq}(X_{S_k}) \end{array} \tag{6.2}$$

where ν is the map of cochains given by

$$\nu(c) = d^*[(c \otimes c \otimes \ldots \otimes c) \otimes_\Gamma 1]. \tag{6.3}$$

Here we have made the identification

$$C^*(X^k_{S_k}) = (C^*(X) \otimes_{\mathbf{Z}} \ldots \otimes_{\mathbf{Z}} C^*(X)) \otimes_\Gamma C^*(A),$$

where $A \to B_{S_k}$ is the universal S_k-bundle and Γ is the integral group ring of S_k, and similarly we identify

$$C^*(X_{S_k}) = C^*(X) \otimes_\Gamma C^*(A).$$

The commutativity of Diagram 6.2 depends of course on the fact that the isomorphism

$$K(X_{2q}, X_{2q-1}) \cong C^{2q}(X)$$

is compatible with (external) products.

The map ν defined by (6.3) induces a map of cohomology (denoted also by ν)

$$\nu: H^{2q}(X, \mathbf{Z}) \to H^{2kq}(X_{S_k}, \mathbf{Z}).$$

The diagrams (6.1) and (6.2) then establish the following

Proposition 6.4. *Let $x \in K_{2q}(X)$ be represented by $a \in H^{2q}(X, \mathbf{Z})$ in the spectral sequence $H^*(X, \mathbf{Z}) \Rightarrow K^*(X)$. Then $\alpha(x^{\otimes k}) \in K_{2kq}(X_{S_k})$ is represented by $\nu(a) \in H^{2kq}(X_{S_k}, \mathbf{Z})$ in the spectral sequence*

$$H^*(X_{S_k}, \mathbf{Z}) \Rightarrow K^*(X_{S_k}),$$

where ν is induced by the formula (6.3).

Remarks. (1) It seems plausible that one could in fact define a tensor-power operation mapping the spectral sequence of X into the spectral sequence of X_{S_k}. Proposition 6.4 concerns itself only with the extreme members E_2 and E_∞ (and only for even dimensions).

(2) The map ν is essentially the parent of all the Steenrod operations, while $x \longmapsto x^{\otimes k}$ is the parent of all the operations in K-theory introduced in § 2. Proposition 6.4 contains therefore, in principle, all the relations between operations in the two theories. We proceed to make this explicit in the simplest case:

Theorem 6.5. *Suppose that X has no torsion so that we may identify $H^*(X, Z_p)$ with $GK^*(X) \otimes Z_p$. If $x \in K_{2q}(X)$ we denote the corresponding element of $H^{2q}(X, Z_p)$ by $\bar{x}$. Let*

$$\psi^p x = \sum_{i=0}^{q} p^{q-i} x_i$$

be the decomposition of $\psi^p x$ given by (5.6). *Then we have*

$$\bar{x}_i = P_p^i(\bar{x}),$$

where
$$P_p^i: H^{2q}(X, Z_p) \to H^{2q+2i(p-1)}(X, Z_p)$$

is the Steenrod power (for $p = 2$ we put $P^i = Sq^{2i}$).

Proof. By Proposition 6.4 the map

$$P: K(X) \to K(X) \otimes R(Z_p)$$

induces
$$\bar{P}: H^*(X, Z_p) \to H^*(X, Z_p) \otimes H^*(Z_p, Z_p), \tag{1}$$

where $\bar{P}$ is ν reduced mod p. Now by (2.6) and Lemma 5.5 (choosing $x_q = x^p$) we have the following expression for $P(x)$,

$$P(x) = x_q \otimes 1 - \sum_{i=0}^{q-1} x_i \otimes p^{q-i-1}(N-p). \tag{2}$$

By definition of the Steenrod powers [(**10**) 112] we have

$$\bar{P}(\bar{x}) = \sum_{i=0}^{q} (-1)^{q-i} P^i(\bar{x}) \otimes \eta^{(q-i)(p-1)},$$

where η is the canonical generator of $H^2(Z_p; Z_p)$.

Comparing (1) and (2) and using Lemma 5.1 we have the result.

Remark. Proposition 6.5, together with the kind of calculations made in (**3**), leads to a very simple proof of the non-existence of elements of Hopf invariant 1 mod p (including the case $p = 2$).

7. Relation with Chern characters

If the space X has no torsion, it is possible to replace the operations ψ^k by the Chern character

$$\mathrm{ch}: K^*(X) \to H^*(X; \mathbf{Q}).$$

In fact ch is a monomorphism and ψ^k can be computed from the formulae

$$\mathrm{ch}\, x = \sum_q \mathrm{ch}_q(x), \qquad x \in K(X),\ \mathrm{ch}_q(x) \in H^{2q}(X; \mathbf{Q})$$

$$\mathrm{ch}\, \psi^k x = \sum_q k^q \, \mathrm{ch}_q(x).$$

Conversely one can define $H^*(X; \mathbf{Q})$ and ch purely in terms of the ψ^k (**3**). It is reasonable therefore to try to express Theorems 5.6 and 6.5 in terms of Chern characters. We shall see that we recover the results of Adams (**1**), at least for spaces without torsion.

If X is without torsion, we identify $H^*(X; \mathbf{Z})$ with its image in $H^*(X; \mathbf{Q})$. If $a \in H^*(X; \mathbf{Q})$, we can write $a = b/d$ for $b \in H^*(X; \mathbf{Z})$ and some integer d. If d can be chosen prime to p, we shall say that a is *p-integral*.

THEOREM 7.1. *Let X be a space without torsion, $x \in K_{2q}(X)$ and p a prime. Then*

$$p^t \, \mathrm{ch}_{q+n}(x)$$

is p-integral, where $t = \left[\dfrac{n}{p-1}\right]$.

Proof. We proceed by induction on n. For $n = 0$ (and all q) the result is a consequence of the periodicity theorem (**8**). We suppose therefore that $n > 0$ and the result established for all $r \leqslant n-1$. By Proposition 5.6 we have

$$\psi^p x = \sum_{i=0}^{q} p^{q-i} x_i, \quad x_i \in K_{2q+2i(p-1)}(X),$$

and so

$$\operatorname{ch} \psi^p x = \sum_{i=0}^{q} p^{q-i} \operatorname{ch} x_i.$$

Taking components in dimension $2(q+n)$ we get

$$p^{q+n} \operatorname{ch}_{q+n}(x) = \sum_{i=0}^{t} p^{q-i} \operatorname{ch}_{q+n}(x_i), \quad t = \left[\frac{n}{p-1}\right]. \tag{1}$$

In particular, for $n = 0$, we have

$$\operatorname{ch}_q(x) = \operatorname{ch}_q(x_0). \tag{2}$$

Since X has no torsion, this implies that

$$y = x_0 - x \in K_{2q+2}(X).$$

Replacing x_0 by $x+y$ in (1) and multiplying by p^{t-q} we get

$$p^t(p^n-1)\operatorname{ch}_{q+n}(x) = p^t \operatorname{ch}_{q+n} y + \sum_{i=1}^{t} p^{t-i} \operatorname{ch}_{q+n}(x_i). \tag{3}$$

But by the inductive hypothesis (with q replaced by $q+1$ and $q+i(p-1)$ $(i \geqslant 1)$) we see that all terms on the right-hand side of (3) are p-integral. Hence $p^t \operatorname{ch}_{q+n}(x)$ is p-integral and so the induction is established.

For any $x \in K_{2q}(X)$ we denote by $\bar{x} \in H^{2q}(X, Z_p)$ the corresponding element obtained from the isomorphism

$$G^{2q}K(X) \otimes Z_p \cong H^{2q}(X; Z_p).$$

Now, by Theorem 7.1, $p^t \operatorname{ch}_{q+t(p-1)} x$ is p-integral. We may therefore reduce it mod p and obtain an element of $H^{2q+2t(p-1)}(X; Z_p)$. It follows from Theorem 7.1 that this depends only on $\bar{x}$. We denote it therefore by $T^t(\bar{x})$, so that T^t is an operation

$$H^{2q}(X; Z_p) \to H^{2q+2t(p-1)}(X; Z_p).$$

We now identify this operation.

Theorem 7.2. *The operation $\sum_{i \geqslant 0} T^i$ is the inverse of the 'total' Steenrod power $\sum_{i \geqslant 0} P^i$, i.e.*

$$(\sum T^i) \circ (\sum P^i) = \text{identity}.$$

Proof. As in Theorem 7.1 we have

$$\psi^p x = \sum_{i=0}^{q} p^{q-i} x_i.$$

Now in equation (1) above take $n = t(p-1)$ and multiply by p^{t-q}. Then reducing mod p we get

$$0 = \sum_{i=0}^{t} T^{t-i}(\bar{x}_i) \quad (t > 0),$$

$$\bar{x} = T^0(\bar{x}_0).$$

But by Theorem 6.5 we have $\bar{x}_i = P^i\bar{x}$, and so we deduce

$$0 = \Big(\sum_{i=0}^{t} T^{t-i}P^i\Big)\bar{x}, \quad \bar{x} = T^0P^0\bar{x}.$$

In other words, the composition

$$(\sum T^i) \circ (\sum P^i)$$

is the identity operator as required.

REFERENCES

1. J. F. Adams, 'On Chern characters and the structure of the unitary group', *Proc. Cambridge Phil. Soc.* 57 (1961) 189–99.
2. ——, 'Vector fields on spheres', *Ann. of Math.* 3 (1962) 603–32.
3. —— and M. F. Atiyah, 'K-theory and the Hopf invariant', *Quart. J. of Math.* (Oxford) (2) 17 (1966) 31–8.
4. M. F. Atiyah, 'Lectures on K-theory' (mimeographed notes, Harvard 1965).
5. ——, 'Characters and cohomology of finite groups', *Publ. Inst. des Hautes Études Sci.* (Paris) 9 (1961) 23–64.
6. ——, 'Vector bundles and the Künneth formula', *Topology* 1 (1962) 245–8.
7. ——, R. Bott, and A. Shapiro, 'Clifford modules', *Topology* 3 (Suppl. 1) (1964) 3–38.
8. —— and F. Hirzebruch, 'Vector bundles and homogeneous spaces', Differential Geometry, Proc. of Symp. in Pure Math. 3 (1961) *American Math. Soc.*
9. —— and G. B. Segal, 'Lectures on equivariant K-theory' (mimeographed notes, Oxford 1965).
10. N. E. Steenrod, 'Cohomology operations', *Ann. of Math. Study* 50 (Princeton, 1962).
11. H. Weyl, *The classical groups* (Princeton, 1949).

Reprinted from Quart. J. Math. (Clarendon Press, Oxford)

K-THEORY AND REALITY

By M. F. ATIYAH

[Received 9 August 1966]

Introduction

THE K-theory of complex vector bundles (**2**, **5**) has many variants and refinements. Thus there are:

(1) K-theory of real vector bundles, denoted by KO,
(2) K-theory of self-conjugate bundles, denoted by KC (**1**) or KSC (**7**),
(3) K-theory of G-vector bundles over G-spaces (**6**), denoted by K_G.

In this paper we introduce a new K-theory denoted by KR which is, in a sense, a mixture of these three. Our definition is motivated partly by analogy with real algebraic geometry and partly by the theory of real elliptic operators. In fact, for a thorough treatment of the index problem for real elliptic operators, our KR-theory is essential. On the other hand, from the purely topological point of view, KR-theory has a number of advantages and there is a strong case for regarding it as the primary theory and obtaining all the others from it. One of the main purposes of this paper is in fact to show how KR-theory leads to an elegant proof of the periodicity theorem for KO-theory, starting essentially from the periodicity theorem for K-theory as proved in (**3**). On the way we also encounter, in a natural manner, the self-conjugate theory and various exact sequences between the different theories. There is here a considerable overlap with the thesis of Anderson (**1**) but, from our new vantage point, the relationship between the various theories is much easier to see.

Recently Karoubi (**8**) has developed an abstract K-theory for suitable categories with involution. Our theory is included in this abstraction but its particular properties are not developed in (**8**), nor is it exploited to simplify the KO-periodicity.

The definition and elementary properties of KR are given in § 1. The periodicity theorem and general cohomology properties for KR are discussed in § 2. Then in § 3 we introduce various derived theories—KR with coefficients in certain spaces—ending up with the periodicity theorem for KO. In § 4 we discuss briefly the relation of KR with Clifford algebras on the lines of (**4**), and in particular we establish a lemma which is used in § 3. The significance of KR-theory for the topological study of real elliptic operators is then briefly discussed in § 5.

This paper is essentially a by-product of the author's joint work with I. M. Singer on the index theorem. Since the results are of independent topological interest it seemed better to publish them on their own.

1. The real category

By a *space with involution* we mean a topological space X together with a homeomorphism $\tau: X \to X$ of period 2 (i.e. $\tau^2 =$ Identity). The involution τ is regarded as part of the structure of X and is frequently omitted if there is no possibility of confusion. A space with involution is just a Z_2-space in the sense of (**6**), where Z_2 is the group of order 2. An alternative terminology which is more suggestive is to call a space with involution a *real* space. This is in analogy with algebraic geometry. In fact if X is the set of complex points of a real algebraic variety it has a natural structure of real space in our sense, the involution being given by complex conjugation. Note that the fixed points are just the real points of the variety X. In conformity with this example we shall frequently write the involution τ as complex conjugation:

$$\tau(x) = \bar{x}.$$

By a *real vector bundle* over the *real space* X we mean a complex vector bundle E over X which is also a real space and such that

(i) the projection $E \to X$ is real (i.e. commutes with the involutions on E, X);

(ii) the map $E_x \to E_{\bar{x}}$ is anti-linear, i.e. the diagram

$$\begin{array}{ccc} \mathbf{C} \times E_x & \to & E_x \\ \downarrow & & \downarrow \\ \mathbf{C} \times E_{\bar{x}} & \to & E_{\bar{x}} \end{array}$$

commutes, where the vertical arrows denote the involution and $\mathbf{C}$ is given its standard real structure $(\tau(z) = \bar{z})$.

It is important to notice the difference between a vector bundle in the category of real spaces (as defined above) and a complex vector bundle in the category of Z_2-spaces. In the definition of the latter the map

$$E_x \to E_{\tau(x)}$$

is assumed to be complex-linear. On the other hand note that if E is a real vector bundle in the category of Z_2-spaces its complexification can be given two different structures, depending on whether

$$E_x \to E_{\tau(x)}$$

is extended linearly or anti-linearly. In the first it would be a bundle in the real category, while in the second it would be a complex bundle in the Z_2-category.

At a fixed point of the involution on X (also called a real point of X) the involution on E gives an anti-linear map

$$\tau_x \colon E_x \to E_x$$

with $\tau_x^2 = 1$. This means that E_x is in a natural way the complexification of a real vector space, namely the $+1$-eigenspace of τ_x (the real points of E_x). In particular if the involution on X is trivial, so that all points of X are real, there is a natural equivalence between the category $\mathscr{E}(X)$ of real vector bundles over X (as space) and the category $\mathscr{F}(X)$ of real vector bundles over X (as real space):† define $\mathscr{E}(X) \to \mathscr{F}(X)$ by $E \mapsto E \otimes_{\mathbf{R}} \mathbf{C}$ ($\mathbf{C}$ being given its standard real structure) and $\mathscr{F}(X) \to \mathscr{E}(X)$ by $F \mapsto F_R$ (F_R being the set of real points of F). This justifies our use of 'real vector bundle' in the category of real spaces: it may be regarded as a natural extension of the notion of real vector bundle in the category of spaces.

If E is a real vector bundle over the real space X then the space $\Gamma(E)$ of cross-sections is a complex vector space with an anti-linear involution: if $s \in \Gamma(E)$, $\bar{s}$ is defined by

$$\bar{s}(x) = \overline{s(\bar{x})}.$$

Thus $\Gamma(E)$ has a real structure, i.e. $\Gamma(E)$ is the complexification of the real vector space $\Gamma(E)_R$.

If E, F are real vector bundles over the real space X a morphism $\phi \colon E \to F$ will be a homomorphism of complex vector bundles commuting with the involutions, i.e.

$$\phi(\bar{e}) = \overline{\phi(e)} \quad (e \in E).$$

$E \otimes_{\mathbf{C}} F$ and $\mathrm{Hom}_{\mathbf{C}}(E, F)$ have natural structures of real vector bundles. For example if $\phi_x \in \mathrm{Hom}_{\mathbf{C}}(E_x, F_x)$ we define $\overline{\phi_x} \in \mathrm{Hom}_{\mathbf{C}}(E_{\bar{x}}, F_{\bar{x}})$ by

$$\overline{\phi_x}(u) = \overline{(\phi_x \bar{u})} \quad (u \in E_{\bar{x}}).$$

It is then clear that a morphism $\phi \colon E \to F$ is just a real section of $\mathrm{Hom}_{\mathbf{C}}(E, F)$, i.e. an element of $(\Gamma\, \mathrm{Hom}_{\mathbf{C}}(E, F))_R$.

If now X is compact then exactly as in (3) [§ 1] we deduce the homotopy property of real vector bundles. The only point to note is that a real section s over a real subspace Y of X can always be extended to a real section over X; in fact if t is any section extending s then $\frac{1}{2}(t+\bar{t})$ is a real extension.

† The morphisms in $\mathscr{F}(X)$ will be defined below.

Suppose now that X is a real algebraic space (i.e. the complex points of a real algebraic variety) then, as we have already remarked, it defines in a natural way a real topological space $X_{\text{alg}} \mapsto X_{\text{top}}$. A *real algebraic vector bundle* can, for our purposes, be taken as a complex algebraic vector bundle $\pi: E \to X$ where X, E, π, and the scalar multiplication $\mathbf{C} \times E \to E$ are all defined over $\mathbf{R}$ (i.e. they are given by equations with real coefficients). Passing to the underlying topological structure it is then clear that E_{top} is a real vector bundle over the real space X_{top}.

Consider as a particular example $X = P(\mathbf{C}^n)$, $(n-1)$-dimensional complex projective space. The standard line-bundle H over $P(\mathbf{C}^n)$ is a real algebraic bundle. In fact H is defined by the exact sequence of vector bundles

$$0 \to E \to X \times \mathbf{C}^n \to H \to 0,$$

where $E \subset X \times \mathbf{C}^n$ consists of all pairs $((z), u) \in X \times \mathbf{C}^n$ satisfying

$$\sum u_i z_i = 0.$$

Since this equation has real coefficients E is a real bundle and this then implies that H is also real. Hence H defines a real bundle over the real space $P(\mathbf{C}^n)$.

As another example consider the affine quadric

$$\sum_{i=1}^{n} z_i^2 + 1 = 0.$$

Since this is affine a real vector bundle may be defined by projective modules over the affine ring $A_+ = \mathbf{R}[z_1, \ldots, z_n]/(\sum z_i^2 + 1)$. Now the intersection of the quadric with the imaginary plane is the sphere

$$\sum_{1}^{n} y_i^2 = 1,$$

the involution being just the anti-podal map $y \mapsto -y$. Thus projective modules over the ring A_+ define real vector bundles over S^{n-1} with the anti-podal involution. If instead we had considered the quadric

$$\sum z_i^2 - 1 = 0$$

then its intersection with the real plane would have been the sphere with trivial involution, so that projective modules over

$$A_- = \frac{\mathbf{R}[z_1, \ldots, z_n]}{(\sum z_i^2 - 1)}$$

define real vector bundles over S^{n-1} with the trivial involution (and so these are real vector bundles in the usual sense). The significance of S^{n-1} in this example is that it is a deformation retract of the quadric in our category (i.e. the retraction preserving the involution).

The Grothendieck group of the category of real vector bundles over a real space X is denoted by $KR(X)$. Restricting to the real points of X we obtain a homomorphism

$$KR(X) \to KR(X_R) \cong KO(X_R).$$

In particular if $X = X_R$ we have

$$KR(X) \cong KO(X).$$

For example taking $X = P(\mathbf{C}^n)$ we have $X_R = P(\mathbf{R}^n)$ and hence a restriction homomorphism

$$KR(P(\mathbf{C}^n)) \to KR(P(\mathbf{R}^n)) = KO(P(\mathbf{R}^n)).$$

Note that the image of $[H]$ in this homomorphism is just the standard real Hopf bundle over $P(\mathbf{R}^n)$.

The tensor product turns $KR(X)$ into a ring in the usual way.

If we ignore the involution on X we obtain a natural homomorphism

$$c\colon KR(X) \to K(X).$$

If $X = X_R$ then this is just complexification. On the other hand if E is a complex vector bundle over X, $E \oplus \tau^*\bar{E}$ has a natural real structure and so we obtain a homomorphism

$$r\colon K(X) \to KR(X).$$

If $X = X_R$ then this is just 'realization', i.e. taking the underlying real space.

2. The periodicity theorem

We come now to the periodicity theorem. Here we shall follow carefully the proof in (**3**) [§ 2] and point out the modifications needed for our present theory.

If E is a real vector bundle over the real space X then $P(E)$, the projective bundle of E, is also a real space. Moreover the standard line-bundle H over $P(E)$ is a real line-bundle. Then the periodicity theorem for KR asserts:

THEOREM 2.1. *Let L be a real line-bundle over the real compact space X, H the standard real line-bundle over the real space $P(L \oplus 1)$. Then, as a $KR(X)$-algebra, $KR(P(L \oplus 1))$ is generated by H, subject to the single relation*

$$([H]-[1])([L][H]-[1]) = 0.$$

First of all we choose a metric in L invariant under the involution. The unit circle bundle S is then a real space. The section z of $\pi^*(L)$ defined by the inclusion $S \to L$ is a *real* section. Hence so are its powers z^k. The isomorphism

$$H^k \cong (1, z^{-k}, L^{-k}) \quad [(3)\ 2.5]$$

is an isomorphism of real bundles. Finally we assert that, if f is a real section of $\operatorname{Hom}(\pi^*E^0, \pi^*E^\infty)$ then its Fourier coefficients a_k are real sections of $\operatorname{Hom}(L^k \otimes E^0, E^\infty)$. In fact we have

$$\bar{a}_k(x) = \overline{a_k(\bar{x})} = -\overline{\frac{1}{2\pi i}\int_{S_{\bar{x}}} f_{\bar{x}} z_{\bar{x}}^{-k-1}\, dz_{\bar{x}}}$$

$$= \frac{1}{2\pi i}\int_{S_x} \overline{f_{\bar{x}}}(\overline{z_{\bar{x}}})^{-k-1}\, d\overline{z_{\bar{x}}} \quad \text{(since the involution reverses the orientation of } S)$$

$$= \frac{1}{2\pi i}\int_{S_x} f_x z_x^{-k-1}\, dz_x \quad \text{(since } f \text{ and } z \text{ are real)}$$

$$= a_k(x).$$

It may be helpful to consider what happens at a real point of X. The condition that f_x is real then becomes

$$f_x(e^{-i\theta}) = \overline{f_x(e^{i\theta})}$$

which implies at once that the Fourier coefficients are real.

Since the linearization procedure of (3) [§ 3] involves only the a_k and and the z^k it follows that the isomorphisms obtained there are all real isomorphisms.

The projection operators Q^0 and Q^∞ of (3) [§ 4] are also real, provided p is real. In fact

$$\bar{Q}_x^0 = \overline{Q_{\bar{x}}^0} = -\overline{\frac{1}{2\pi i}\int_{S_{\bar{x}}} p_{\bar{x}}^{-1}\, dp_{\bar{x}}}$$

$$= \frac{1}{2\pi i}\int_{S_x} (\overline{p_{\bar{x}}})^{-1}\, d\overline{p_{\bar{x}}}$$

$$= \frac{1}{2\pi i}\int_{S_x} p_x^{-1}\, dp_x, \qquad \text{since } p \text{ is real.}$$

Similarly for Q^∞. The bundle $V_n(E^0, p, E^\infty)$ is therefore real and (4.6) is an equation in $KR(P)$. The proof in § 5 now applies quite formally.

We are now in a position to develop the usual cohomology-type theory, using relative groups and suspensions. There is, however, one new feature here which is important. Besides the usual suspension, based on $\mathbf{R}$ with trivial involution, we can also consider $\mathbf{R}$ with the involution $x \mapsto -x$. It is often convenient to regard the first case as the real axis $\mathbf{R} \subset \mathbf{C}$ and the second as the imaginary axis $i\mathbf{R} \subset \mathbf{C}$, the complex numbers $\mathbf{C}$ always having the standard real structure given by complex conjugation. We use the following notation:

$$R^{p,q} = \mathbf{R}^q \oplus i\mathbf{R}^p,$$

$$B^{p,q} = \text{unit ball in } R^{p,q},$$

$$S^{p,q} = \text{unit sphere in } R^{p,q}.$$

Note that $R^{p,p} \cong \mathbf{C}^p$. Note also that, with this notation, $S^{p,q}$ has dimension $p+q-1$.

The relative group $KR(X,Y)$ is defined in the usual way as $\widetilde{KR}(X/Y)$ where $\widetilde{KR}$ is the kernel of the restriction to base point. We then define the (p,q) suspension groups

$$KR^{p,q}(X,Y) = KR(X\times B^{p,q}, X\times S^{p,q} \cup Y\times B^{p,q}).$$

Thus the usual suspension groups KR^{-q} are given by

$$KR^{-q} = KR^{0,q}.$$

As in (2) one then obtains the exact sequence for a real pair (X,Y)

$$\ldots \to KR^{-1}(X) \to KR^{-1}(Y) \to KR(X,Y) \to KR(X) \to KR(Y). \quad (2.2)$$

Similarly one has the exact sequence of a real triple (X,Y,Z). Taking the triple $(X\times B^{p,0},\ X\times S^{p,0}\cup Y\times B^{p,0},\ X\times S^{p,0})$ one then obtains an exact sequence

$$\ldots \to KR^{p,1}(X) \to KR^{p,1}(Y) \to KR^{p,0}(X,Y) \to KR^{p,0}(X) \to KR^{p,0}(Y)$$

for each integer $p \geqslant 0$.

The ring structure of $KR(X)$ extends in a natural way to give external products

$$KR^{p,q}(X,Y) \otimes KR^{p',q'}(X',Y') \to KR^{p+p',q+q'}(X'',Y''),$$

where $X'' = X\times X'$, $Y'' = X\times Y' \cup X'\times Y$. By restriction to the diagonal these define internal products.

We can reformulate Theorem 2.1 in the usual way. Thus let

$$b = [H]-1 \in KR^{1,1}(\text{point}) = KR(B^{1,1}, S^{1,1}) = KR(P(\mathbf{C}^2))$$

and denote by β the homomorphism

$$KR^{p,q}(X,Y) \to KR^{p+1,q+1}(X,Y)$$

given by $x \mapsto b.x$. Then we have

THEOREM 2.3. *β: $KR^{p,q}(X, Y) \to KR^{p+1,q+1}(X, Y)$ is an isomorphism.*

Note also that the exact sequence of a real pair is compatible with the periodicity isomorphism. Hence if we define

$$KR^p(X,Y) = KR^{p,0}(X,Y) \quad \text{for } p \geqslant 0$$

it follows that the exact sequence (2.2) for (X, Y) can be extended to infinity in both directions. Moreover we have natural isomorphisms $KR^{p,q} \cong KR^{p-q}$.

We consider now the general Thom isomorphism theorem as proved for K-theory in (**2**) [§ 2.7]. We recall that the main steps in the proof proceed as follows:

(i) for a line-bundle we use (2.1),
(ii) for a decomposable vector bundle we proceed by induction using (2.1),
(iii) for a general vector bundle we use the splitting principle.

An examination of the proof in (**2**) [§ 2.7] shows that the only point requiring essential modification is the assertion that a vector bundle is locally trivial and hence locally decomposable. Now a real vector bundle has been defined as a vector bundle with a real structure. Thus it has been assumed locally trivial as a vector bundle in the category of spaces. What we have to show is that it is also *locally trivial in the category of real spaces*. To do this we have to consider two cases.

(i) $x \in X$ a real point. Then $E_x \cong \mathbf{C}^n$ in our category. Hence by the extension lemma there exists a real neighbourhood U of x such that $E|U \cong U \times \mathbf{C}^n$ in the category.
(ii) $x \neq \bar{x}$. Take a complex isomorphism $E_x \cong \mathbf{C}^n$. This induces an isomorphism $E_{\bar{x}} \cong \mathbf{C}^n$. Hence we have a real isomorphism
$$E|Y \cong Y \times \mathbf{C}^n,$$
where $Y = \{x, \bar{x}\}$. By the extension lemma there exists a real neighbourhood U of Y so that $E|U \cong U \times \mathbf{C}^n$.

Thus we have

THEOREM 2.4 (Thom Isomorphism Theorem). *Let E be a real vector bundle over the real compact space X. Then*
$$\phi \colon KR(X) \to \widetilde{KR}(X^E)$$
is an isomorphism where $\phi(x) = \lambda_E . x$ and λ_E is the element of $\widetilde{KR}(X^E)$ defined by the exterior algebra of E.

Among other results of (**2**) [§ 2.7] we note the following:
$$\begin{aligned} KR(X \times P(\mathbf{C}^n)) &\cong KR(X)[t]/t^n - 1 \\ &\cong KR(X) \otimes_{\mathbf{Z}} K(P(\mathbf{C}^n)). \end{aligned}$$

We leave the computation of KR for Grassmannians and Flag manifolds as exercises for the reader. The determination of KR for quadrics is a more interesting problem, since the answer will depend on the signature of the quadratic form.

We conclude with the following observation. Consider the inclusion

$$R^{0,1} = \mathbf{R} \xrightarrow{i} \mathbf{C} = R^{1,1}.$$

This induces a homomorphism

$$\begin{array}{ccc} K^{1,1}(\text{point}) & \xrightarrow{i^*} & K^{0,1}(\text{point}) \\ \| & & \| \\ \widetilde{KR}(P(\mathbf{C}^2)) & \to & \widetilde{KR}(P(\mathbf{R}^2)). \end{array}$$

Since $i^*[H]$ is the real Hopf bundle over $P(\mathbf{R}^2)$ it follows that $\eta = i^*(b) = i^*([H]-1)$ is the reduced Hopf bundle over $P(\mathbf{R}^2)$.

3. Coefficient theories

If Y is a fixed real space then the functor $X \mapsto KR(X \times Y)$ gives a new cohomology theory on the category of real spaces which may be called *KR-theory with coefficients in Y*. We shall take for Y the spheres $S^{p,0}$ (where the involution is the anti-podal map). A theory F will be said to have period q if we have a natural isomorphism $F \cong F^{-q}$. Then we have.

Proposition 3.1. *KR-theory with coefficients in $S^{p,0}$ has period*

$$\begin{aligned} &2 \textit{ if } p = 1, \\ &4 \textit{ if } p = 2, \\ &8 \textit{ if } p = 4. \end{aligned}$$

Proof. Consider R^p as one of the three fields $\mathbf{R}$, $\mathbf{C}$, or $\mathbf{H}$ ($p = 1, 2,$ or 4). Then for any real space X the map

$$\mu_p : X \times S^{p,0} \times R^{0,p} \to X \times S^{p,0} \times R^{p,0}$$

given by $\mu_p(x, s, u) = (x, s, su)$, where su is the product in the field, is a real isomorphism. Hence it induces an isomorphism

$$\mu_p^* : KR^{p,0}(X \times S^{p,0}) \to KR^{0,p}(X \times S^{p,0}).$$

Replacing X by a suspension gives an isomorphism

$$\mu_p^* : KR^{p,q}(X \times S^{p,0}) \to KR^{0,p+q}(X \times S^{p,0}).$$

Taking $q = p$ and using the isomorphism

$$\beta^p : KR \to KR^{p,p}$$

given by Theorem 2.1, we obtain finally an isomorphism

$$\begin{array}{rcl} \mu_p^* \beta^p : KR(X \times S^{p,0}) & \to & KR^{0,2p}(X \times S^{p,0}) \\ & & \| \\ & & KR^{-2p}(X \times S^{p,0}). \end{array}$$

Remark. μ^* is clearly a $KR(X)$-module homomorphism. Since the same is true of β this implies that the periodicity isomorphism

$$\gamma_p = \mu_p^* \beta^p : KR(X \times S^{p,0}) \to KR^{-2p}(X \times S^{p,0})$$

is multiplication by the image c_p of 1 in the isomorphism

$$KR(S^{p,0}) \to KR^{-2p}(S^{p,0}).$$

This element c_p is given by

$$c_p = \gamma_p(1) = \mu^*(b^p . 1), \quad 1 \in KR(S^{p,0}).$$

For any Y the projection $X \times Y \to X$ will give rise to an exact coefficient sequence involving KR and KR with coefficients in Y. When Y is a sphere we get a type of Gysin sequence:

PROPOSITION 3.2. *The projection* $\pi : S^{p,0} \to$ point *induces the following exact sequence*

$$\dots \to KR^{p-q}(X) \xrightarrow{\chi} KR^{-q}(X) \xrightarrow{\pi^*} KR^{-q}(X \times S^{p,0}) \xrightarrow{\delta} \dots$$

where χ *is the product with* $(-\eta)^p$, *and* $\eta \in KR^{-1}(\text{point}) \cong \widetilde{KR}(P(\mathbf{R}^2))$ *is the reduced real Hopf bundle.*

Proof. We replace π by the equivalent inclusion $S^{p,0} \to B^{p,0}$. The relative group is then $KR^{p,q}(X)$. To compute χ we use the commutative diagram

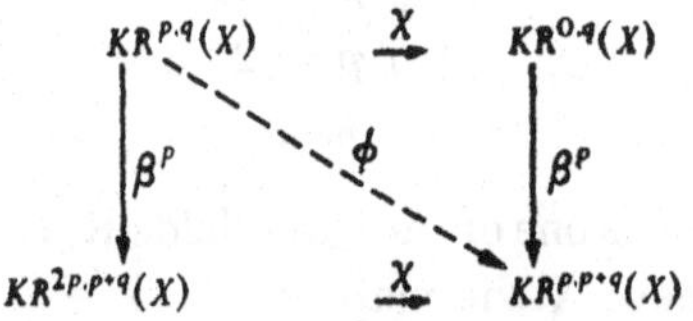

Let θ be the automorphism of $K^{2p,p+q}(X)$ obtained by interchanging the two factors $R^{p,0}$ which occur. Then the composition $\chi\theta\beta^p$ is just multiplication by the image of b^p in

$$KR^{p,p}(\text{point}) \to KR^{0,p}(\text{point}).$$

But this is just η^p. It remains then to calculate θ. But the usual proof given in (2) [§ 2.4] shows that $\theta = (-1)^{p^2} = (-1)^p$.

We proceed to consider in more detail each of the theories in (3.1). For $p = 1$, $S^{p,0}$ is just a pair of conjugate points $\{+1, -1\}$. A real vector bundle E over $X \times \{+1, -1\}$ is entirely determined by the complex vector bundle E_+ which is its restriction to $X \times \{+1\}$. Thus we have

PROPOSITION 3.3. *There is a natural isomorphism*

$$KR(X \times S^{1,0}) \cong K(X).$$

Note in particular that this does not depend on the real structure of X but just on the underlying space. The period 2 given by (3.1) confirms what we know about $K(X)$. The exact sequence of (3.2) becomes now

$$\ldots \to KR^{1-q}(X) \xrightarrow{\chi} KR^{-q}(X) \xrightarrow{\pi^*} K^{-q}(X) \xrightarrow{\delta} KR^{2-q}(X) \to \ldots \qquad (3.4)$$

where χ is multiplication by $-\eta$ and $\pi^* = c$ is complexification. We leave the identification of δ as an exercise for the reader. This exact sequence is well-known (when the involution on X is trivial) but it is always deduced from the periodicity theorem for the orthogonal group. Our procedure has been different and we could in fact use (3.4) to prove the orthogonal periodicity. Instead we shall deduce this more easily later from the case $p = 4$ of (3.1).

Next we consider $p = 2$ in (3.1). Then $KR^{-q}(X \times S^{2,0})$ has period 4. We propose to identify this with a self-conjugate theory. If X is a real space with involution τ a *self-conjugate* bundle over X will mean a complex vector bundle E together with an isomorphism $\alpha: E \to \overline{\tau^* E}$. Consider now the space $X \times S^{2,0}$ and decompose $S^{2,0}$ into two halves $S^{2,0}_+$ and $S^{2,0}_-$ with intersection $\{\pm 1\}$.

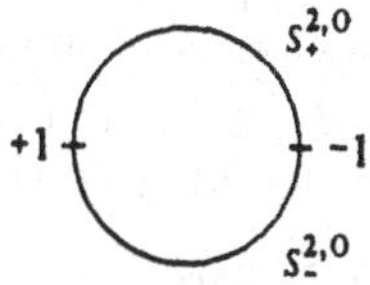

It is clear that to give a real vector bundle F over $X \times S^{2,0}$ is equivalent to giving a complex vector bundle F_+ over $X \times S^{2,0}_+$ (the restriction of F) together with an isomorphism

$$\phi: F|X \times \{+1\} \to \tau^*(\overline{F}|X \times \{-1\}).$$

But $X \times \{+1\}$ is a deformation retract of $X \times S^{2,0}_+$ and so [cf. (3) 2.3] we have an isomorphism

$$\theta: F_+|X \times \{-1\} \to F_+|X \times \{+1\}$$

unique up to homotopy. Thus to give ϕ is equivalent, up to homotopy, to giving an isomorphism

$$\alpha: E \to \overline{\tau^* E},$$

where E is the bundle over X induced from F_+ by $x \mapsto (x, 1)$ and

$$\alpha_x = \theta_{(\bar{x},-1)} \phi_{(x,1)}.$$

In other words *isomorphism classes of real bundles over $X \times S^{2,0}$ correspond bijectively to homotopy classes of self-conjugate bundles over X.* Moreover this correspondence is clearly compatible with tensor products.

Now let $KSC(X)$ denote the Grothendieck group of homotopy classes of self-conjugate bundles over X. If τ is trivial this agrees with the definitions of (1) and (7). Then we have established

PROPOSITION 3.5. *There is a natural isomorphism of rings*

$$KSC(X) \to KR(X \times S^{2,0}).$$

The exact sequence of (3.2), with $p = 2$, then gives an exact sequence

$$\ldots \to KR^{2-q}(X) \xrightarrow{\chi} KR^{-q}(X) \xrightarrow{\pi^*} KSC^{-q}(X) \xrightarrow{\delta} KR^{3-q}(X) \to \ldots \quad (3.6)$$

where χ is multiplication by η^2 and π^* is the map which assigns to any real bundle the associated self-conjugate bundle (take $\alpha = \tau$). The periodicity in KSC is given by multiplication by a generator of $KSC^{-4}(\text{point})$.

Finally we come to the case $p = 4$. For this we need

LEMMA 3.7. *Let $\eta \in KR^{-1}(\text{point})$ be the element defined in § 2. Then $\eta^3 = 0$.*

Proof. This can be proved by linear algebra. In fact we recall [(4) § 11] the existence of a homomorphism $\alpha: A_k \to KR^{-k}(\text{point})$ where the A_k are the groups defined by use of Clifford algebras. Then η is the image of the generator of $A_1 \cong Z_2$ and $A_3 = 0$. Since the homomorphisms α_k are multiplicative [(4) § 11.4] this implies that $\eta^3 = 0$.

COROLLARY 3.8. *For any $p \geqslant 3$ we have short exact sequences*

$$0 \to KR^{-q}(X) \xrightarrow{\pi^*} KR^{-q}(X \times S^{p,0}) \xrightarrow{\delta} KR^{p+1-q}(X) \to 0.$$

Proof. This follows from (3.7) and (3.2).

According to the remark following (3.1) the periodicity for $KR(X \times S^{4,0})$ is given by multiplication with the element

$$c_4 = \mu_4^*(b^4 . 1) \in KR^{-8}(S^{4,0}).$$

Now recall [(4) Table 2] that $A_8 \cong \mathbf{Z}$, generated by an element λ (representing one of the irreducible graded modules for the Clifford algebra C_8). Applying the homomorphism

$$\alpha: A_8 \to KR^{-8}(\text{point})$$

we obtain an element $\alpha(\lambda) \in KR^{-8}(\text{point})$. The connexion between c_4 and $\alpha(\lambda)$ is then given by the following lemma:

LEMMA 3.9. *Let 1 denote the identity of $KR(S^{4,0})$. Then*

$$c_4 = \alpha(\lambda) . 1 \in KR^{-8}(S^{4,0}).$$

The proof of (3.9) involves a careful consideration of Clifford algebras and is therefore postponed until § 4 where we shall be discussing Clifford algebras in more detail.

Using (3.9) we are now ready to establish

THEOREM 3.10. *Let $\lambda \in A_8$, $\alpha(\lambda) \in KR^{-8}(\text{point})$ be as above. Then multiplication by $\alpha(\lambda)$ induces an isomorphism*

$$KR(X) \to KR^{-8}(X)$$

Proof. Multiplying the exact sequence of (3.8) by $\alpha(\lambda)$ we get a commutative diagram of exact sequences

$$\begin{array}{ccccccccc} 0 \to & KR^{-q}(X) & \to & KR^{-q}(X \times S^{4,0}) & \to & KR^{5-q}(X) & \to 0 \\ & \downarrow \phi_q & & \downarrow \psi_q & & \downarrow \phi_{5-q} & \\ 0 \to & KR^{-q-8}(X) & \to & KR^{-q-8}(X \times S^{4,0}) & \to & KR^{-3-q}(X) & \to 0. \end{array}$$

By (3.9) we know that ψ_q coincides with the periodicity *isomorphism* γ_4. Hence ϕ_q is a monomorphism for all q. Hence ϕ_{5-q} in the above diagram is a monomorphism, and this, together with the fact that ψ_q is an isomorphism, implies that ϕ_q is an epimorphism. Thus ϕ_q is an isomorphism as required.

Remark. If the involution on X is trivial, so that $KR(X) = KO(X)$, this is the usual 'real periodicity theorem'.

By considering the various inclusions $S^{q,0} \to S^{p,0}$ we obtain interesting exact sequences. For the identification of the relative group we need

LEMMA 3.11. *The real space (with base point) $S^{p,0}/S^{q,0}$ is isomorphic to*

$$S^{p-q,0} \times B^{q,0}/S^{p-q,0} \times S^{q,0}.$$

Proof. $S^{p,0} - S^{q,0}$ is isomorphic to $S^{p-q,0} \times R^{q,0}$. Now compactify.

COROLLARY 3.12. *We have natural isomorphisms:*

$$KR(X \times S^{p,0},\ X \times S^{q,0}) \cong KR^{0,q}(X \times S^{p-q,0}).$$

In view of (3.8) the only interesting cases are for low values of p, q. Of particular interest is the case $p = 2$, $q = 1$. This gives the exact sequence [cf. (**1**)]

$$\ldots \to K^{-1}(X) \to KSC(X) \to K(X) \to K(X) \to \ldots .$$

The exact sequence of (3.8) does in fact split canonically, so that (for $p \geqslant 3$)

$$KR^{-q}(X \times S^{p,0}) \cong KR^{-q}(X) \oplus KR^{p+1-q}(X). \qquad (3.13)$$

To prove this it is sufficient to consider the case $p = 3$, because the general case then follows from the commutative diagram ($p \geqslant 4$)

$$\begin{array}{ccccc} 0 \to & KR(X) & \to & KR(X \times S^{p,0}) \\ & \downarrow & & \downarrow \\ 0 \to & KR(X) & \to & KR(X \times S^{3,0}) \end{array}$$

obtained by restriction. Now $S^{3,0}$ is the 2-sphere with the anti-podal involution and this may be regarded as the conic $\sum_0^2 z_i^2 = 0$ in $P(\mathbf{C}^3)$. In § 5 we shall give, without proof, a general proposition which will imply that, when Y is a quadric,

$$KR(X) \to KR(X \times Y)$$

has a canonical left inverse. This will establish (3.13).

4. Relation with Clifford algebras

Let Cliff($R^{p,q}$) denote the Clifford algebra (over $\mathbf{R}$) of the quadratic form

$$-\left(\sum_1^p y_i^2 + \sum_1^q x_j^2\right)$$

on $R^{p,q}$. The involution $(y, x) \mapsto (-y, x)$ of $R^{p,q}$ induces an involutory automorphism of Cliff($R^{p,q}$) denoted by† $a \mapsto \bar{a}$.

Let $M = M^0 \oplus M^1$ be a complex Z_2-graded Cliff($R^{p,q}$)-module. We shall say that M is a *real* Z_2-graded Cliff($R^{p,q}$)-module if M has a real structure (i.e. an anti-linear involution $m \mapsto \bar{m}$) such that

(i) the Z_2-grading is compatible with the real structure, i.e.

$$\bar{M}^i = M^i \quad (i = 0, 1),$$

(ii) $\overline{am} = \bar{a}\bar{m}$ for $a \in$ Cliff($R^{p,q}$) and $m \in M$.

Note that if $p = 0$, so that the involution on Cliff($R^{p,q}$) is trivial, then

$$M_R = M_R^0 \oplus M_R^1 = \{m \in M \mid \bar{m} = m\}$$

is a real Z_2-graded module for the Clifford algebra in the usual sense [a C_q-module in the notation of (**4**)].

The basic construction of (**4**) carries over to this new situation. Thus a real graded Cliff($R^{p,q}$)-module $M = M^0 \oplus M^1$ defines a triple (M^0, M^1, σ) where $\sigma\colon S^{p,q} \times M^0 \to S^{p,q} \times M^1$ is a real isomorphism given by

$$\sigma(s, m) = (s, sm).$$

In this way we obtain a homomorphism

$$h\colon M(p,q) \to KR^{p,q}(\text{point})$$

where $M(p,q)$ is the Grothendieck group of real graded Cliff($R^{p,q}$)-modules. If M is the restriction of a Cliff($R^{p,q+1}$)-module then σ extends over $S^{p,q+1}$. Since the projection

$$S_+^{p,q+1} \to B^{p,q}$$

† This notation diverges from that of (**4**) [§ 1] where (for $q = 0$) this involution is called α and 'bar' is reserved for an anti-automorphism.

is an isomorphism of real spaces (S_+ denotes the upper hemisphere with respect to the last coordinate) it follows that M defines the zero element of $KR^{p,q}(\text{point})$. Hence, defining $A(p,q)$ as the cokernel of the restriction

$$M(p,q+1) \to M(p,q),$$

we see that h induces a homomorphism

$$\alpha: A(p,q) \to KR^{p,q}(\text{point}).$$

Moreover, as in (4), α is multiplicative. Note that for $p = 0$ this α coincides essentially with that defined in (4), since

$$A(0,q) \cong A_q,$$
$$KR^{0,q}(\text{point}) \cong KO^{-q}(\text{point}).$$

The exterior algebra $\Lambda^*(\mathbf{C}^1)$ defines in a natural way a Cliff$(R^{1,1})$-module by

$$z(1) = ze, \quad z(e) = -\bar{z}1$$

where $1 \in \Lambda^0(\mathbf{C}^1)$ and $e \in \Lambda^1(\mathbf{C}^1)$ are the standard generators. Let $\lambda_1 \in A(1,1)$ denote the element defined by this module. In view of the definition of $b \in KR^{1,1}(\text{point})$ we see that

$$\alpha(\lambda_1) = -b$$

and hence, since α is multiplicative,

$$\alpha(\lambda_1^4) = b^4.$$

Let M be a graded Cliff$(R^{4,4})$-module representing λ_1^4 (in fact as shown in (4) [§ 11], we can construct M out of the exterior algebra $\Lambda^*(\mathbf{C}^4)$), and let $w = e_1 e_2 e_3 e_4 \in$ Cliff$(R^{4,4})$ where e_1, e_2, e_3, e_4 are the standard basis of $R^{4,0}$. Then we have

$$w^2 = 1, \quad \bar{w} = w,$$
$$wz = \bar{z}w \quad \text{for } z \in \mathbf{C}^4 = R^{4,4}.$$

Hence we may define a new anti-linear involution $m \mapsto \tilde{m}$ on M by

$$\tilde{m} = -w\bar{m}$$

and we have

$$\widetilde{zm} = -w\overline{zm} = -w\bar{z}\bar{m} = -zw\bar{m}$$
$$= z\tilde{m}.$$

Thus M with this new involution (or real structure) is a real graded Cliff$(R^{0,8})$-module, a C_8-module in the notation of (4): as such we denote it by N. From dimensional considerations [cf. (4) Table 2], we see that it must be one of the two irreducible C_8-modules. But on complexification (i.e. ignoring involutions) it gives the same as M and hence N represents the element of A_8 denoted in (4) by λ.

After these preliminaries we can now proceed to the proof of Lemma 3.9. What we have to show is that under the map

$$\mu_4 : S^{4,0} \times \mathbf{R}^8 \to S^{4,0} \times \mathbf{C}^4$$

the element of $KR^{4,4}(S^{4,0})$ defined by M lifts to the element of $KR^{-8}(S^{4,0})$ defined by N. To do this it is clearly sufficient to exhibit a commutative diagram of real isomorphisms

$$\begin{array}{ccc} S^{4,0} \times \mathbf{R}^8 \times N & \xrightarrow{\nu} & S^{4,0} \times \mathbf{C}^4 \times M \\ \downarrow & & \downarrow \\ S^{4,0} \times \mathbf{R}^8 \times N & \xrightarrow{\nu} & S^{4,0} \times \mathbf{C}^4 \times M \end{array} \tag{4.1}$$

where ν is compatible with μ_4 (i.e. $\nu(s, x, y, n) = (s, x+isy, m)$ for some m) and the vertical arrows are given by the module structures (i.e. $(s, x, y, n) \mapsto (s, x, y, (x, y)n)$.

Consider now the algebra $\text{Cliff}(R^{4,0}) = C_4$. The even part C_4^0 is isomorphic to $\mathbf{H} \oplus \mathbf{H}$ [(4) Table 1]. Moreover its centre is generated by 1 and $w = e_1 e_2 e_3 e_4$, the two projections being $\frac{1}{2}(1 \pm w)$. To be quite specific let us define the embedding

$$\xi : \mathbf{H} \to \text{Cliff}^0(R^{4,0})$$

by

$$\xi(1) = \frac{1+w}{2},$$

$$\xi(i) = \frac{1+w}{2} e_1 e_2,$$

$$\xi(j) = \frac{1+w}{2} e_1 e_3,$$

$$\xi(k) = \frac{1+w}{2} e_1 e_4.$$

Then we can define an embedding

$$\eta : S(\mathbf{H}) \to \text{Spin}(4) \subset \Gamma_4$$

by $\eta(s) = \xi(s) + \frac{1}{2}(1-w)$, where Γ_4 is the Clifford group [(4) 3.1] and $S(\mathbf{H})$ denotes the quaternions of norm 1. It can now be verified that the composite homomorphism

$$S(\mathbf{H}) \to \text{Spin}(4) \to SO(4)$$

defines the natural action of $S(\mathbf{H})$ on $\mathbf{R}^4 = \mathbf{H}$ given by left multiplication.† In other words

$$\eta(s) y \eta(s)^{-1} = sy \qquad (s \in S(\mathbf{H}),\ y \in \mathbf{R}^4). \tag{4.2}$$

If we give $S(\mathbf{H})$ the anti-podal involution then η is *not* compatible with involutions, since the involution on the even part C_4^0 is trivial.

† We identify $1, i, j, k$ with the standard base e_1, e_2, e_3, e_4 in that order.

Regarding Cliff($R^{4,0}$) as embedded in Cliff($R^{4,4}$) in the natural way we now define the required map ν by

$$\nu(s, x, y, n) = (s, x+isy, \eta(s)n).$$

From the definition of w it follows that

$$\eta(s)w = -\eta(-s)$$

and so $\quad \eta(-s)\bar{n} = \eta(-s)\{-w\bar{n}\} = \eta(s)\bar{n} = \overline{\eta(s)n},$

showing that ν is a *real* map. Equation (4.2) implies that

$$\eta(s)(x, y)n = (x+isy)\eta(s)n,$$

showing that ν is compatible with the module structures. Thus we have established the existence of the diagram (4.1) and this completes the proof of Lemma 3.9.

The definitions of $M(p,q)$ and $A(p,q)$ given were the natural ones from our present point of view. However, it may be worth pointing out what they correspond to in more concrete or classical terms. To see this we observe that if M is a real $C(R^{p,q})$-module we can define a new action [ ] of R^{p+q} on M by

$$[x, y]m = xm+iym.$$

Then $$[x, y]^2m = \{-\|x\|^2+\|y\|^2\}m.$$

Moreover for the involutions we have

$$\begin{aligned} \overline{[x, y]m} &= \overline{xm}+\overline{iym} \\ &= x\bar{m}+iy\bar{m} \quad (\text{since } \bar{y} = -y) \\ &= [x, y]\bar{m}. \end{aligned}$$

Thus M_R is now a real module in the usual sense for the Clifford algebra $C_{p,q}$ of the quadratic form

$$Q(p,q) \equiv \sum_1^p y_i^2 - \sum_1^q x_j^2.$$

It is easy to see that we can reverse the process. *Thus $M(p,q)$ can equally well be defined as the Grothendieck group of real graded $C_{p,q}$-modules.* From this it is not difficult to compute the groups $A_{(p,q)}$ on the lines of (**4**) [§ 4, 5] and to see that they depend only on $p-q$ (mod 8) [cf. also (**8**)]. Using the result of (**4**) [11.4] one can then deduce that

$$\alpha\colon A(p,q) \to KR^{p,q}(\text{point})$$

is always an isomorphism. The details are left to the reader. We should perhaps point out at this stage that our double index notation was suggested by the work of Karoubi (**8**).

The map α can be defined more generally for principal spin bundles as in (**4**) and we obtain a Thom isomorphism theorem for spin bundles on the lines of (**4**) [12.3]. We leave the formulation to the reader.

5. Relation with the index

If $\hat{\phi}$ denotes the Fourier transform of a function ϕ then we have

$$\hat{\bar{\phi}}(x) = \overline{\hat{\phi}(-x)}.$$

Since the symbol $\sigma(P)$ of an elliptic differential operator P is defined by Fourier transforms (**9**) it follows that

$$\sigma(\bar{P})(x, \xi) = \overline{\sigma(P)(x, -\xi)}$$

where $\bar{P}$ is the operator defined by

$$\bar{P}\phi = \overline{P\bar{\phi}}.$$

Here we have assumed that P acts on functions so that $\overline{P\bar{\phi}}$ is defined. More generally if X is a *real differentiable manifold*, i.e. a differentiable manifold with a differentiable involution $x \mapsto \bar{x}$, and if E, F are real differentiable vector bundles over X, then the spaces $\Gamma(E)$, $\Gamma(F)$ of smooth sections have a real structure and for any linear operator

$$P: \Gamma(E) \to \Gamma(F)$$

we can define $\bar{P}: \Gamma(E) \to \Gamma(F)$ by

$$\bar{P}(\phi) = \overline{P\bar{\phi}}.$$

If P is an elliptic differential operator then

$$\sigma(\bar{P})(x, \xi) = \overline{\sigma(P)(\bar{x}, -\tau^*(\xi))}. \tag{5.1}$$

It is natural to define P to be a *real operator* if $P = \bar{P}$. If the involution on X is trivial this means that P is a differential operator with real coefficients with respect to real local bases of E, F. In any case it follows from (5.1) that the symbol $\sigma(P)$ of a real elliptic operator gives an isomorphism of real vector bundles

$$\pi^*E \to \pi^*F,$$

where $\pi: S(X) \to X$ is the projection of the cotangent sphere bundle and we define the involution on $S(X)$ by

$$(x, \xi) \to (\bar{x}, -\tau^*(\xi)).$$

Note that if τ is the identity involution on X the involution on $S(X)$ is not the identity but is *the anti-podal map on each fibre*. This is the basic reason why our KR-theory is needed here. In fact the triple

$$(\pi^*E, \pi^*F, \sigma(P))$$

defines in the usual way an element

$$[\sigma(P)] \in KR(B(X), S(X))$$

where $B(X)$, the unit ball bundle of $S(X)$, has the associated real structure.†

The kernel and cokernel of a real elliptic operator have natural real structures. Thus the index is naturally an element of KR(point). Of course since

$$KR(\text{point}) \to K(\text{point})$$

is an isomorphism there is no immediate advantage in defining this apparently refined real index. However, the situation alters if we consider instead a *family* of real elliptic operators with parameter or base space Y. In this case a real index can be defined as an element of $KR(Y)$ and

$$KR(Y) \to K(Y)$$

is not in general injective.

All these matters admit a natural extension to real elliptic complexes (**9**). Of particular interest is the Dolbeault complex on a real algebraic manifold. This is a real elliptic complex because the holomorphic map $\tau: X \to \bar{X}$ maps the Dolbeault complex of $\bar{X}$ into the Dolbeault complex of X. If X is such that the sheaf cohomology groups $H^q(X, \mathcal{O}) = 0$ for $q \geqslant 1$, $H^0(X, \mathcal{O}) \cong \mathbf{C}$, the index, or Euler characteristic, of the Dolbeault complex is 1. Based on this fact one can prove the following result:

PROPOSITION. *Let $f: X \to Y$ be a fibering by real algebraic manifolds, where the fibre F is such that*

$$H^q(F, \mathcal{O}) = 0 \qquad (q \geqslant 1,\ H^0(F, \mathcal{O}) \cong \mathbf{C}),$$

then there is a homomorphism

$$f_*: KR(X) \to KR(Y)$$

which is a left inverse of

$$f^*: KR(Y) \to KR(X).$$

The proof cannot be given here but we observe that a special case is given by taking $X = Y \times F$ where F is a (compact) homogeneous space of a real algebraic linear group. For example we can take F to be a complex quadric, as required to prove (3.13). We can also take $F = SO(2n)/U(n)$, or $SO(2n)/T^n$, the flag manifold of $SO(2n)$. These spaces can be used to establish the splitting principle for orthogonal bundles. It is then significant to observe that the real space

$$\{SO(2n)/U(n)\} \times R^{0,2n}$$

† All this extends of course to integral (or pseudo-differential) operators.

has the structure of a real vector bundle. A point of $SO(2n)/U(n)$ defines a complex structure of R^{2n} and conjugate points give conjugate structures. For $n = 2$ this is essentially† what we used in § 3 to deduce the orthogonal periodicity from Theorem 2.1.

† In (3.1) we used the 3-sphere $S^{4,0}$. We could just as well have used the 2-sphere $S^{3,0}$. This coincides with $SO(4)/U(2)$.

REFERENCES

1. D. W. Anderson (Thesis: not yet published).
2. M. F. Atiyah, Lectures on *K*-theory (mimeographed notes, Harvard 1965).
3. M. F. Atiyah and R. Bott, 'On the periodicity theorem for complex vector bundles', *Acta Math.* 112 (1964) 229–47.
4. M. F. Atiyah, R. Bott, and A. Shapiro, 'Clifford modules', *Topology* 3 (1964) 3–38.
5. M. F. Atiyah and F. Hirzebruch, 'Vector bundles and homogeneous spaces', *Proc. Symposium in Pure Math.* Vol. 3, American Mathematical Society (1961).
6. M. F. Atiyah and G. B. Segal, 'Equivariant *K*-theory' (Lecture notes, Oxford 1965).
7. P. S. Green, 'A cohomology theory based upon self-conjugacies of complex vector bundles, *Bull. Amer. Math. Soc.* 70 (1964) 522.
8. M. Karoubi (Thesis: not yet published).
9. R. Palais, 'The Atiyah–Singer index theorem', *Annals. of Math.* Study 57 (1965).

The Mathematical Institute
Oxford University

Printed in the United States
by Baker & Taylor Publisher Services

Printed in the United States
by Baker & Taylor Publisher Services